Owen, Bruce M.

Video economics.

$35.00

DATE			

Video Economics

Video Economics

Bruce M. Owen
Steven S. Wildman

Harvard University Press
Cambridge, Massachusetts
London, England
1992

Printed in the United States of America
10 9 8 7 6 5 4 3 2 1

This book is printed on acid-free paper.

Library of Congress Cataloging-in-Publication Data

Owen, Bruce M.
Video economics / Bruce M. Owen and Steven S. Wildman.
p. cm.
Includes bibliographical references and index.
ISBN 0-674-93716-3
1. Television broadcasting—United States. I. Wildman, Steven S.
II. Title.
HE8700.8.093 1992
384.55'1-dc20 91-32924
CIP

To the memory of my grandfather, James Brandon Manning.
BMO

To my parents, Rex and Lorabel Wildman.
SSW

Contents

Preface

Powerful economic forces are bringing about fundamental change in the U.S. video industry. In the 1990s this industry includes a wide array of organizations that distribute video programming to the home—in particular, broadcast stations, cable television systems, and the networks with which they are affiliated. Some of the industry's elements have been part of television's traditional structure since the 1950s and are now undergoing dramatic alteration; others, such as pay-per-view cable and the video cassette rental market, are new developments. This book is intended for those who must understand the economics of the video industry in order to work successfully in it: students of broadcasting, communication, and journalism; advertising and media executives and planners; and government policymakers and their staffs.

In analyzing television markets, we make two assumptions. The first concerns the positive value of economic efficiency, which has a meaning different from technical or business efficiency in the colloquial sense. As economists, we are interested in the functioning of markets—in this case, television markets. Inefficiency in these markets concerns us because it injures the public. A market can be inefficient in at least three ways. First, firms may consistently fail to minimize the cost of producing their output (or, equivalently, fail to maximize output with given resources). Second, firms may fail to produce the type and quantity of output that maximize consumer welfare. Third, constraints, usually inspired by government, may be present that prevent the sale of certain products or services or the very existence of a market in these services. Generally speaking, markets that are competitive tend to be efficient, whereas markets characterized by significant monopoly power tend to be inefficient. (We leave aside natural monopolies—situations in which costs can be minimized only if one firm supplies 100 percent of the market. As we shall see in Chapter 6, cable television firms may have this characteristic in a local area, depending on how the market is defined.) In the television market, where consumers most often do not pay directly for the services they consume, it is often difficult to make strong statements about the relative efficiency of alternative market structures.

Our second assumption concerns the First Amendment. We believe that government intrusion in the television medium is to be avoided, and that monopoly in television is especially undesirable. The First Amendment may seem perfectly clear: "Congress shall make no law . . . abridging the freedom of speech, or of the press." But the Supreme Court has upheld a number of federal laws abridging the freedom of broadcasters.

Economic efficiency and the spirit of the First Amendment sometimes conflict, but usually they reinforce each other. And frequently in pursuing one, broadcasters and policymakers can more nearly attain the other.

From its inception until the late 1970s, broadcast regulation was caught in a conflict of its own design. The government created broadcasters with monopoly power and obliged them to exercise control over program content and to offer unprofitable program services in the public interest. The government then proceeded to regulate, more or less directly, the broadcasters' exercise of market power. This approach grew out of a series of judicial and legislative decisions that consciously rejected both the antipaternalistic ethic of the First Amendment and the market as an allocative device (Spitzer, 1986; Powe, 1987; Hazlett, 1990a), and is now widely acknowledged by regulators and scholars alike to have been misguided. Nevertheless, many would still argue that television is inherently different from the other media—more powerful, influential, or insidious—and that this McLuhanesque distinction justifies or compels government regulation, regardless of the structure of the medium. Such observers put television regulation in the same category as laws regulating the safety of food and drugs. We want to make it plain that our differences with this approach are philosophical (and logical), not "economic."

A common rejoinder to dicta concerning the constitutional role of the press is to point to the actual content of the media and to ask whether it is worthy of constitutional protection. Although the network news may make a valuable contribution to the marketplace of ideas, "Wheel of Fortune" has little obvious relationship to libertarian ideals. Indeed, most of television is entertainment, not ideas in any conventional sense. Is the dichotomy between entertainment and news (or opinion) a valuable one for policy purposes? This is dangerous ground for two reasons. First, some entertainment contains important ideological content. Second, the consumption of ideas and information cannot be easily distinguished from the consumption of entertainment. Ideas and information are often consumed because they are packaged in an entertaining way.

Much of the content of that common whipping boy, the commercial, is information—sometimes valuable information. Although television commercials typically contain less "straight" information than do newspaper want ads or space advertising in print media, they contain some. The role of television, like that of other media, is to broadcast information because information is a commodity in demand and thus attracts audiences. Television entertainment programming is very popular because it is free, or at least cheap, and because it packages information in an engaging way. That the content of the information is often trivial or banal is not of any particular significance from a constitutional point of view.

The role of the media is to provide conduits for the flow of information and expression in society. The content or packaging of that information is separable from the degree of efficiency and freedom with which the media operate. This is true even though one cannot deny that the "medium is the message." Surely the economic structure and the technology of the media condition the nature, content, and effect of messages. This is what is meant by the relationship of media structure to efficiency and freedom.

We believe that television should be structured in a way that is the most competitive and that generates the greatest freedom of expression and consumption of ideas and information. The distinction between efficiency and freedom becomes particularly important when there are trade-offs between the efficiency of information production or consumption and the distribution of the information, or when freedom for producers and consumers of information can be augmented only at the expense of the owners of the media.

Our backgrounds go beyond the purely academic. One of us has had federal government experience; both of us have been consultants for media firms, including broadcasters, broadcast networks, cable operators, and related trade associations. These experiences have helped us to make this book more relevant to "real world" concerns than it otherwise might have been. Our hope is that the reader of this book will come away prepared to think more deeply about the business strategy and government policy goals determined by video economics.

The support of the Ameritech Foundation is gratefully acknowledged. We are also grateful for comments and suggestions from Mike Baumann, James Dertouzos, James Ettema, Mark Frankena, Peter Greenhalgh, Jonathan Levy, Phil Nelson, Joel Rosenbloom, George Vradenburg III, and James Webster. Harold Furchtgott-Roth and Kent Mikkelsen made major contributions to Chapter 6, including much of the econometric work. Portions of Chapter 3 are based on the corresponding chapter of *Television Economics,* by Owen, J. H. Beebe, and W. G. Manning, Jr. (Lexington, Mass.: Lexington Books, 1974). Lauren Feinswog, Wendy Learmont, Nancy Lee, B. J. Macatulad, Dan Merkle, and Leigh Roveda provided skilled research and editorial assistance. Michael Aronson provided persistent encouragement without which this book would not exist. Barbara de Boinville and Elizabeth Gretz were creative and helpful editors. No one but the authors is responsible for the remaining errors.

June 1991

Video Economics

Abbreviations

ABC	American Broadcasting Companies
ATRC	Advanced Television Research Committee
ATV	Advanced television
CBS	Columbia Broadcasting System
CNN	Cable News Network
DBS	Direct broadcast satellite
EDTV	Enhanced definition television
ESPN	Entertainment and Sports Programming Network
FCC	Federal Communications Commission
HBO	Home Box Office
HDTV	High definition television
HHI	Hirschman-Herfindahl Index
IDTV	Improved definition television
MDS	Multipoint distribution system
MMDS	Multipoint multichannel distribution service
MSO	Multiple (cable) system operator
NAB	National Association of Broadcasters
NBC	National Broadcasting Company
NTSC	National Television Standards Committee
OAR	Open architecture receiver
PBS	Public Broadcasting Service
PPV	Pay per view
SMATV	Satellite master antenna television
TCI	Telecommunications Inc.
VCR	Video cassette recorder

1 / Introduction

Few industries have been as revolutionized by regulatory reform and technological change as the television business. In the early 1970s, "television" meant broadcast television. Observers of the industry were preoccupied with the scarcity of television channels, the economic consequences for broadcasters and consumers, and the misguided federal policies that had created and perpetuated that scarcity. In the 1990s, channels abound. "Television" means not just broadcast television, which transmits signals over the air from local stations to household antennas, but cable television, which uses a wire (a coaxial cable) to deliver signals to the home; video cassette television; and direct satellite-to-home television. In the early 1970s, when the president's press secretary called in "the networks" for an announcement, he contacted ABC, CBS, and NBC. Today a call also goes to CNN, which broadcasts news 24 hours a day. In the seventies, for a typical viewer who decided to spend the evening watching television, the choices were ABC, CBS, or NBC. Today that viewer can watch four commercial broadcast networks (including Fox), several independent broadcast stations, dozens of national cable networks, and hundreds of movies and other programs available in local retail outlets on prerecorded rental video cassettes. Television, with the exception of PBS, was once entirely advertiser supported. Today most of the viewing options require a viewer payment, and much of the programming has no commercials. Television has become more like the highly competitive, pluralistic magazine industry than the tight-knit oligopoly of television's own first few decades. To mark this change, we refer to the "video" industry rather than to the television industry.

This book focuses on the fundamental economics of television in the 1990s. An earlier work, Owen, Beebe, and Manning (1974), principally addressed policy issues, because at that time the structure of the television industry was determined by federal regulatory policy, and the performance of the industry was a matter of deep concern to federal policymakers. Here we are interested primarily in the strategic problems of firms that compete in the video business and only secondarily in policy analysis. Today the structure of the industry is shaped largely by market forces, and policymakers' concern with the performance of the industry is less intense than it was. Nevertheless, important regulatory and policy issues remain. They range from the antitrust and franchising policy toward cable television to the continuing political battles to reintroduce the fairness doctrine and to regulate program content.

Broadcasting, moreover, continues to be an industry in which important gains over rivals can be made through the intervention of federal regulators. Lobbyists representing Hollywood business interests know this well. Making use of, and protecting against others' use of, Washington regulators is a central element of business success, and understanding the policy analysis of television issues is as important for industry executives as for academic analysts.[1]

Glamour and social influence notwithstanding, television is a business. Television performance is primarily influenced by the economic incentives of owners, managers, customers, suppliers, and employees and by the structure of television markets. Noneconomic goals and values are not unimportant in the industry, but policy debate about television too often ignores precisely those economic factors that can shed the most light on the merits of alternative policies, or that most plausibly explain existing behavior.

This chapter presents a general framework for economic analysis of television markets, together with some technical and historical background. The regulatory reform movement is also discussed. Chapter 2 deals with the program supply market, a market in which competition and ease of entry play an important role. The quantity of programs supplied is quite responsive to changes in the demand for them. Chapter 2 also discusses "windowing," the practice of staggering release dates to various media in order to exploit opportunities to discriminate in price. In Chapter 3 we explore traditional models of program choice under various parametric conditions, including advertiser demand, viewer preferences, and market structure. The chapter examines how market structure affects producers' program strategies and viewer satisfaction. A more modern perspective on program choice is adopted in Chapter 4, which assesses the implications of market structure for consumer welfare. Chapter 5 analyzes the role and behavior of television networks, both broadcast and cable. Regulatory reform and new technology have revolutionized the major broadcast networks and the strategic options they face. Cable television, the first technology to offer television of abundance, is discussed in Chapter 6. Chapter 7, on "advanced" or high definition television, examines the policy and strategic choices available to the federal government and individual segments of the video industry.

Among the topics not considered in detail are copyright economics and exclusivity, sports broadcasting, the political and social effects of programming content, public television, and spectrum allocation reform. Many of these issues, however, are mentioned in passing, and the analysis in the book is in large part applicable to their study.

Basic Economics of Broadcast Television

The first and most serious mistake that an analyst of the television industry can make is to assume that advertising-supported television broadcasters are in business to broadcast programs. They are not. Broadcasters are in the business of producing *audiences*. These audiences, or means of access to them, are sold to advertisers. The product of a television station is measured in dimensions of people and time. The price of the product is quoted in dollars per thousand viewers per unit of commercial time, typically 20 or 30 seconds.

It is often said that television stations seek the largest possible audiences, but this is an oversimplification. First, advertisers are interested not merely in the size of an audience but in its characteristics. In the trade these characteristics are called "demographics," and they refer to the age, sex, and income composition of the audience. Some audiences of a given size are thus more valuable than others. Second, a television station may be able to increase its audience only at a prohibitive program cost. If station managers are rational, as their stockholders expect them to be, they will be interested in maximizing the difference between advertising revenue and costs. This difference, of course, is profit. Although it is true that stations are interested in achieving as large an audience as possible for any given program expenditure, they do not seek to obtain an indefinitely large audience regardless of the cost.

Why do television stations sell access to audiences instead of to programs? Traditionally, most stations have been unable to collect revenue directly from viewers. Until the mid-1970s it was *illegal,* under FCC rules, to charge viewers for most television programs.[2] Moreover, the nature of the medium complicates such efforts. It is technically difficult and expensive, for example, to scramble signals and rent decoders, as satellite networks now attempt to do in order to collect from individuals with backyard dishes. A more fundamental reason for the sale of audiences to advertisers is that audiences are valuable to advertisers. Audiences that pay something to receive messages (as with magazines and newspapers) are often even more valuable to advertisers than audiences generated by free messages, because it is clear that the paying audience is interested. Eventually, commercials of one kind or another will almost certainly be presented with many pay programs, just as some movie theaters now insert commercials with their previews and some retail video cassettes or movies contain commercials. Indeed, "basic" cable television service today is really nothing more than multichannel pay television with commercials.

Because there is competition among television stations and between the stations and other media, individual stations must attempt to increase their long-run profits. But if this is true, why do they broadcast some programs that draw audiences smaller than those that could be obtained with other programming at the same cost? Examples include public affairs programs, documentaries, presidential speeches, congressional hearings, and some religious shows. Owners of a station can choose to "give up" otherwise attainable profits in order to do public service programming. To the extent that they do so, they may simply be deciding on one of many ways in which to spend their income. But a more plausible explanation of such programming lies in regulation, community relations, and advertiser preferences. Television stations operate under five-year federal licenses that can be revoked if the Federal Communications Commission finds that a station has not been operating in the "public interest." Although television licenses are seldom revoked, the threat is nevertheless real. A license is a valuable asset and is not lightly risked. The FCC associates certain kinds of programs (namely, those listed above) with performance in the public interest. Special interest groups have successfully demanded such programs as the price of their support for license renewal. If it appears to the licensee that this price is less than the cost of legal service—plus the risk of license revocation times the value of the license—then the station will acquiesce. Even today, in a more relaxed federal regulatory atmosphere, stations must consider their relationships with local government and community leaders and with advertisers.

The economic motivation of television stations and other video enterprises is taken for granted here. We assume that licensees want to maximize their long-run profits, taking due account of regulation, the behavior of their competitors, risk-avoidance, and so on. The paradigm of economic behavior is useful in understanding television markets, and equally useful in strategic and policy analysis. We do not regard it as inherently either "bad" or "good" that television stations, networks, cable systems, and video specialty stores seek profits. Rather, we examine the consequences of the motivation for strategic and policy choices.

Players in the Industry

Local retail outlets—television stations, cable systems, and to a lesser extent video specialty stores—are the backbone of the commercial television system. Many other groups and institutions are also important. Most prominent are the networks, which act as economic agents or brokers for local outlets both in selling audiences and in acquiring programs. Networks are the distributors or wholesalers of the television product. The original manufacturers, of course, are the Hollywood program and movie producers. The federal government intervenes in the relationships among these groups in important ways, often to benefit one at the expense of another. These and other players are the cast of this book (see

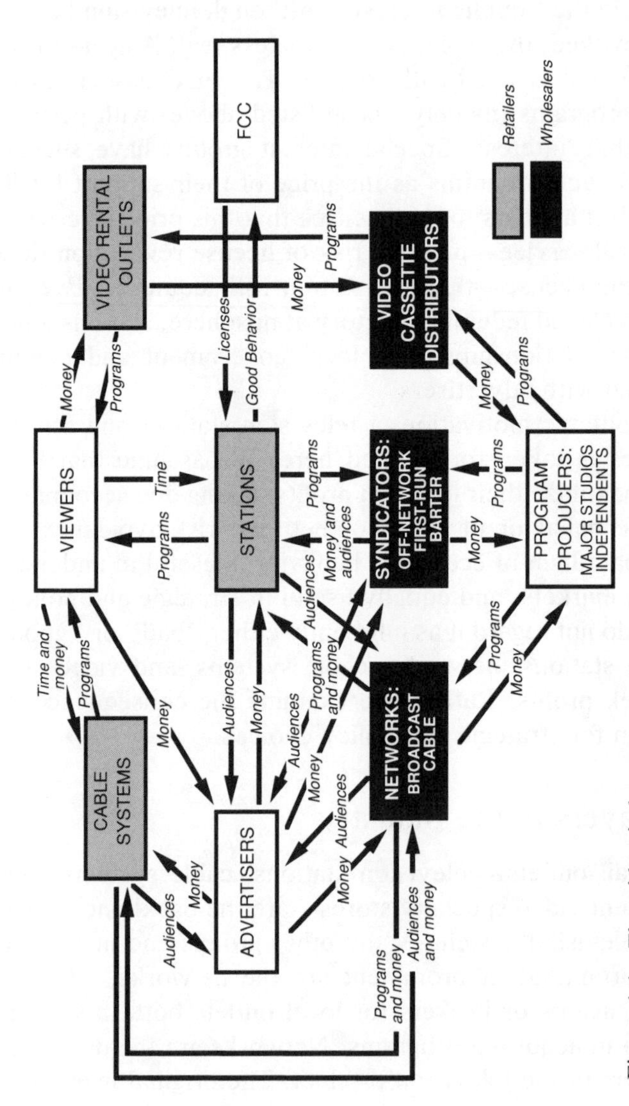

Figure 1.1 The video industry

Figure 1.1). Their interactions are sometimes strange and complex, but never uninteresting.

Stations

There are about 1,000 full-power commercial television stations in the United States, and more than 300 educational or public broadcasting stations. In addition, there are more than 2,000 "low-power" stations on the air or under construction. Stations can be broken down into very high frequency (VHF) and ultra high frequency (UHF) categories (see Table 1.1).

The approximately 650 stations, both VHF and UHF, affiliated with the three major broadcast networks, ABC, NBC, and CBS, are the largest single group in terms of numbers, revenues, and profits, and historically these stations have attracted the most viewers. But independent stations (not affiliated with one of the three major broadcast networks) have grown rapidly in number and importance in recent years; they purchase their programming directly from Hollywood without relying on a network intermediary. (Fox, however, is trying to become an intermediary for many of these stations.) And UHF stations, once the stepchildren of the industry because of reception problems, are becoming increasingly competitive. Cable subscribers can receive UHF signals equivalent in quality to VHF signals, because UHF signals transmitted over wires do not attenuate as they do over the air. In the last decade,

Table 1.1 Television broadcast stations

Type	VHF	UHF
In operation		
Commercial	552	563
Noncommercial	125	229
Low power	185	594
Under construction		
Commercial	18	184
Noncommercial	3	16
Low power	165	1,056

Source: Broadcasting, February 11, 1991, p. 79.

the FCC granted construction permits for several hundred new UHF stations.

The FCC initiated a program in 1980 to permit the construction of thousands of low-power television stations. These stations are permitted to broadcast on the same frequencies as stations in nearby cities, but at power levels low enough not to cause interference. About 600 low-power stations were on the air in 1990, about 200 of them commercial stations, and the FCC has been besieged with applications for licenses.

Cable Systems

There are more than 10,000 local cable television systems in the United States, many of them independently owned. Cable systems typically enjoy a de facto monopoly right to provide cable service in a local area. The vast majority (over 80 percent) of cable systems have at least 30 channels. Newer ones have more than 100 channels and can provide such interactive services as special movies and sports events on demand to individual households. Chains of cable systems (called multiple system operators, or MSOs) are increasing in size. The largest, TeleCommunications Inc. (TCI), in 1990 had interests in systems serving nearly a quarter of all subscribers. Many of the major MSOs are vertically integrated, owning interests in one or more national cable networks.

Networks

ABC, CBS, and NBC shared less than 60 percent of the prime-time television audience in 1990, down from 90 percent in 1980. In the intervening decade these three commercial broadcast networks were joined by a fourth, Fox, and by many part-time or ad hoc broadcast networks and dozens of cable networks.

Cable networks are of two kinds: basic and pay. Basic networks are generally offered by cable systems in a bundle, which includes local broadcast signals for a lump-sum monthly fee. Pay networks are usually offered by the cable system as separate services for an additional fee. Some pay networks offer viewers specific events on a pay-per-view (PPV) basis. The largest basic networks are Entertainment and Sports Programming Network, with access to 53 million cable households; Cable News Network, 52 million; and USA Network (general entertainment), 50 million. The largest pay networks are Home Box Office (movies), with 17 million subscribing households; Showtime (movies), 7 million; Cinemax (movies, owned by HBO), 6 million; and the Disney Channel (family), 4 million. Basic networks rely on advertising revenue and payments from cable systems. A few basic networks pay cable systems for access.

The networks serve as intermediaries between local stations or cable systems and both advertisers and program suppliers. The three major broadcast networks, once the most important players in the television industry, are today in a painful decline caused by increased competition and federal regulations that still restrict their strategic options (see Chapter 5).

Video Outlets

In the 1990s more than two-thirds of all households with a television own a video cassette recorder, and thousands of video specialty outlets rent or sell prerecorded cassettes. VCR viewing now accounts for a significant fraction of television viewing. Wholesale distribution of prerecorded video cassettes, once dominated by independent concerns such as Vestron, is increasingly controlled by the major Hollywood studios.

Other Distribution Media

There are a variety of other distribution media, none of them very important in the early 1990s, but each a potential source of competition for conventional means of distribution. Multipoint, multichannel distribution systems—so-called wireless cable—provide from half a dozen to as many as twenty channels of coordinated pay programming to about one million households. Many firms offer a form of localized cable service known as SMATV (satellite master antenna television). Plans to provide direct broadcast satellite (DBS) service have been announced by several entities. Several DBS services may be operating by the mid-1990s.

Viewers

Ninety-two million U.S. households have television sets, and most have more than one. Nearly all have color sets. About 60 percent of all television households are cable subscribers, and about 90 percent could become cable subscribers because a cable runs nearby. Two-thirds have at least one VCR. A few million, mainly rural, households (no one is sure exactly how many) own backyard satellite antennas that can receive network signals intended for cable head ends or broadcast stations. Such households can pay a monthly fee to receive these network signals; satellite distribution companies, such as CSBN and Netlink, now provide subscription services for these viewers. Advertiser-supported broadcasters and cable networks depend on audience measurement services, such as Arbitron and A. C. Nielsen, to convince advertisers that viewers are watching. (Cable operators know how many subscribers send in their monthly checks and how many viewers use special pay-per-view services.)

In recent years the broadcast networks, which pay Nielsen some $10 million a year to measure audiences, have been at odds with Nielsen over "people meter" measurements. These meters are attached to a sample of television sets to record what viewers are watching and when. With these tallies, combined with demographic information about the sample households and more detailed data from diaries kept by other sample viewing households, Nielsen projects viewing levels according to certain age, sex, education, and income categories. People meters record viewing by requiring each individual viewer to respond affirmatively when queried. Audiences measured in this way are typically smaller than are audiences measured by telephone surveys, meters attached passively to sets, or diaries. In addition, the networks contend that Nielsen undercounts certain populations, such as those on vacation or traveling.[3]

In contrast to the issue of how *many* people are watching television, there is little mystery about *how* people watch. Research has long emphasized the passivity of viewers. Viewing is based more on the availability of spare time than on the availability of preferred content (Webster and Wakshlag, 1983). "The main gratification from TV viewing is simply watching, regardless of content, and it is only as a more or less secondary consideration that program choices are made'' (Glick and Levy, 1962, p. 115). A major study of human behavior with respect to television viewing (which concluded, inter alia, that television viewing is addictive) did not consider program content of sufficient relevance even to include it as an explanatory variable (Kubey and Csikszentmihalyi, 1990, pp. 79–80). Apparently, human brain wave activity, while differing between television and print, does not differ significantly across program content. New wide-screen, high definition television, however, ''may involve greater cognitive activity than did traditional television viewing'' (ibid., p. 137).

Advertisers

Advertisers seek to reach potential purchasers of their products, and these purchasers are not distributed uniformly across the population. Accordingly, each advertiser will have a different set of preferences for audiences. The most prominent of these preferences are geography and demographic composition (age, sex, income, and education). Many advertisers are concerned with the timing of their advertising campaigns; they need to coordinate exposure in different media and to coordinate advertising with retail promotions and with production and inventory activities.

There are two distinct advertising marketplaces: national advertising and local advertising. Advertisers seeking to reach the entire national audience have a choice of media: broadcast television networks, cable networks, magazines, and newspaper inserts. Television advertising accounts for only about 22 percent of all advertising expenditure. National network advertising on television is sold at a high price (more than $400,000 per 30-second commercial in prime time for the major networks) because it reaches the largest audiences that can be obtained through any medium. But it is also sold at a relatively low price per viewer (about 10¢ per viewing home per hour) because it is not very selective. Cable networks and so-called barter-syndication program producers, discussed below, also sell national television advertising (see Table 1.2).

National advertisers can reach particular local audiences by purchasing time directly from local television stations or their repre-

Table 1.2 Television advertising expenditures, 1990 (in millions of dollars)

Advertiser	Expenditure
National	
Broadcast network	9,617
Cable network	1,816
Barter syndication	1,557
National spot	7,715
Total	20,705
Local	
Broadcast	7,912
Cable	635
Total	8,547
Total	29,252

Source: Kagan Media Index, February 28, 1991, pp. 12–13.

sentatives. This is called national spot advertising. Advertisers interested in this market can choose among television stations, radio stations, newspapers, regional editions of national magazines, and regional feeds of broadcast networks. Whether a given national advertiser uses network or spot ads (or in what proportion) depends on cost/reach trade-offs.

Local advertising is bought by firms or merchants with only local sales who want to reach local audiences, and these firms deal directly with the station. National spot advertising is distinguished from local advertising by the identity of the buyers. In the national spot market, big advertisers deal with local stations' representatives or with program producers. The latter bundle together ads and programs, then offer the package to local stations with much of the commercial time presold to national advertisers. This practice, known as "barter syndication," has grown rapidly in recent years, as has cable network advertising.

Spot advertising sales bypass the networks. The advantage to advertisers is that the ad campaign can be tailored to the geographic audience most likely to purchase the product. Although the cost per thousand viewers is theoretically lower (because of transaction costs) for a network campaign than for an equivalent set of national spot campaigns, the network buy generally involves some wasted expense: the cost of exposure to audiences in portions of the country where the advertiser has few outlets or otherwise poor prospects. From this perspective, network sales and national spot sales are best regarded as differentiated products in the same market. Advertisers generally use both, with the proportion of their expenditures adjusted to reflect, among other things, the relative prices of these two means of advertising.

Direct viewer payment. Advertising once was the sole source of television revenue. But since 1977, when the D.C. Circuit Court of Appeals struck down the FCC's restrictions on pay television in its *Home Box Office* opinion, 567 F.2d 9 (D.C.Cir. 1977), revenue from viewers has become increasingly important. All cable television is in part viewer supported. Some subscription services for over-the-air television also exist, although their future is unclear. Today direct viewer payments for television service account for about one-third of total revenues of the television industry, and this fraction will grow in the 1990s. Table 1.3 provides a rough overview of

Table 1.3 Monthly per household media expenditures by consumers and advertisers, 1990

Type of media	Monthly expenditure
Video cassettes[a]	$12.79
Cable television[b]	25.85
Television advertising[c]	
Broadcast network	8.61
Spot	6.91
Local	7.08
Motion pictures (box office gross)[c]	4.59
Newspapers[c]	
Circulation revenue	7.79
Advertising revenue	27.28
Magazines[c]	
Circulation revenue	8.16
Advertising revenue	8.39
Radio advertising[c]	
Spot	1.43
Local	5.98

Source: Based on estimates by the *Kagan Media Index,* February 28, 1991, pp. 12–13.

a. Sales and rentals; expenditure per VCR household.

b. Basic and premium service; expenditure per cable household.

c. Expenditure per television household.

the various media revenue sources, expressed on a monthly per household basis.

The competition. The television industry, broadcast as well as cable, competes for advertising with other media, notably radio broadcasting, newspapers, and magazines. Each of these media offers special advantages to certain types of advertisers, depending on their product, the amount and kind of information that is customarily transmitted in their advertisements, and the geographical dispersion and identifiability of their potential customers. Television is entirely unsuitable for some advertisers; for others it is almost indispensable. But most advertisers can substitute one medium for another in response to changes in prices of advertising time or space. Competition between television stations or networks and other media for advertising dollars may be nearly as fierce as competition among television outlets.

Because advertising demand depends on audience size and demographics, the media also compete for audiences. Over the years, such intermedia competition has proved to be extremely important. The growth of television has caused losses in newspaper readership, mass magazine circulation, radio listening, and especially—at least until recently—movie theater attendance. Cable television, in turn, has similarly begun to "fragment" the television audience.

The television industry competes for consumer entertainment dollars as well as for advertising dollars. Consumers seeking video entertainment have numerous choices. They can rent or buy a video cassette, go to the movies, buy cable service (basic or premium), or watch "free" over-the-air television.

Regulators

Television would be a more ordinary business if it did not use the electromagnetic spectrum to transmit its signals. An act of Congress nationalized the broadcast spectrum in the United States in 1912, and ever since then use of the spectrum has been allocated by the federal government. Today television stations use the spectrum under a renewable five-year license from the Federal Communications Commission. For years the FCC exercised its licensing power in such a way that most of the activities of television stations, except advertising prices, were directly regulated. Even advertising prices were indirectly regulated by virtue of the FCC's power to control entry and the number of competitors.[4] Despite this regulation, use of the spectrum is not essential to the television business, which can use cable or prerecorded media, and regulation of program content is not essential to spectrum allocation.

Only a relatively small part of the usable electromagnetic spectrum is used for commercial broadcasting. Radar, taxicab radios, cellular telephone services, satellites, microwave systems, citizens' band radios, and police, fire, and military communication systems all employ the spectrum. These users are constantly clamoring for more spectrum, which is free and therefore cheaper than investments in equipment to accommodate more intensive use of a given spectrum allocation. Because the FCC is charged with allocating the spectrum in the "public interest," it is faced with the difficult task of allocating an economic resource without the usual market signals from prices. The spectrum could probably be bought and sold as private property rights (Coase, 1959; DeVany et al., 1969). If it were, and if the property rights were sufficiently well defined so that significant interference problems were avoided, the public would be better assured that the spectrum was used in the most productive manner. But the public has no such assurance. That portion of the spectrum allocated to television, from an economic point of view, is of arbitrary size.

VHF-TV licenses have been relatively scarce. This scarcity has led to economic rents—profits in excess of those required to keep stations in business—that accrue to the holders of licenses (Greenberg, 1969; Levin, 1971b; Webbink, 1973). Lord Thompson of Fleet, a British press baron, supposedly once characterized a television station as a "license to print money." In competitive industries, entry by new firms takes place until excess profits are reduced to zero, and only sufficient profits are earned to return the market price of capital and other inputs. For many years FCC decisions about spectrum allocation for television prevented entry from reducing VHF profits in this way. Consequently, a television license is a valuable asset that commands a considerable price in the market. License holders are willing to pay a considerable price to proceed in a way that satisfies either the regulatory objectives of the Commission or the licensees' potential challengers at license-renewal time. There is an active market in licenses, and some firms even specialize in the brokerage of television and radio stations. Permission for such sales must be obtained from the FCC, but the transfer process is usually pro forma.

Until about 1976 the quasi-official view of broadcasting at the FCC was that broadcasters were public-spirited citizens, fiduciaries of the public, who were unfortunately obliged to sell advertising in order to defray expenses of operation. The main responsibility of broadcasters was to the viewing public, and not to shareholders, in the view of the FCC. This unrealistic view was the source of many ineffective policies and two considerable evils. The first was that broadcasters took advantage of the public interest myth to promote protectionist policies motivated in fact by economic self-interest. The second evil was that the public was misled in its perception of the role and function of broadcasting in the United States.

The artificial scarcity of VHF-TV licenses created a system of powerful vested interests that stood in the path of reform and change—particularly change that increased competition and viewers' choices. For many years the FCC behaved as if it were the host at an exclusive dinner party. Only a few, cherished, upper-crust guests were invited, gaining profit from their exclusivity, while the riffraff were firmly shut out. How did all this come about, and why and to what extent has it changed?

Regulatory Reform

Just after World War II the FCC created a system of television broadcasting that continued the concentration of economic power that had already evolved in radio broadcasting. (Hazlett, 1990a, provides a comprehensive review of the early history of broadcast regulation.) The FCC decided that the public interest would be better served by providing each viewer with more local signals than by providing greater overall choice. The result was a system in which most cities had only three television stations, and the nation therefore could have only three television networks. These licenses were distributed broadly across congressional districts. The artificial scarcity of channels created a windfall flow of profits to the stations and to the networks. For more than three decades, the

FCC defended these economic interests against encroachments from new technology and other forms of competition.

As with any oligarchical system, a myth of social responsibility evolved; the myth served to rationalize the privileges of the wealthy few. Broadcasters and networks, together with the FCC, evolved a code of "fiduciary" responsibility to the public, embodied in FCC regulations and in the policies of the National Association of Broadcasters. Under the code the broadcasters performed such public services as providing news, public affairs, and religious programming. These good works were regarded as a sufficient moral justification for the broadcasters' privileged economic position. Any competitive threat to that position presumably threatened the public's assumed interest in unprofitable public service programming and thus justified the government's legal barriers to entry into broadcasting. The FCC, pressed by the courts, evolved a legal principle, known as the Carroll doctrine, which said that the FCC must decide whether the economic effect of a new broadcast license in a given area would damage existing service. The prevention of financial damage to existing broadcasters thus became part of the public interest obligation of the federal government.

By 1954 it had become apparent that advertising revenues could support many more television channels than had been authorized. Nevertheless, the FCC permitted only weak and ineffective UHF competitors into the local markets. In 1958 the FCC could have remedied the scarcity of channels by making each city either purely UHF or purely VHF. This would have eliminated the engineering and image disadvantages of stations on UHF. The FCC, however, failed to act.[5] In 1962, when cable television first began to threaten broadcasters' economic interests, the FCC asserted jurisdiction over it. The Commission found the likelihood of cable's "adverse impact upon potential and existing service . . . too substantial to be dismissed" (Ginsburg, 1979, p. 337).[6] Accordingly, in 1966, 1968, and 1971, motivated in part by the Supreme Court's permissive interpretation of the copyright obligations of cable operators, the FCC imposed highly restrictive regulations on cable that were designed to prevent harm to broadcasters.[7] After 1966 the FCC similarly restricted pay television. It also tried to prevent increases in the prices broadcasters would have to pay for programs.[8]

In short, the FCC acted as if it were the servant and protector of the broadcast industry. Although clothed in public interest rhetoric, most regulations served the economic interests of the VHF broadcasters and networks.[9] As the FCC later tacitly admitted, these policies made broadcasters rich at the expense of the viewing public.

Impetus for Reform

With the exception of pioneering work by Coase (1950, 1959, 1962), there was little academic interest in broadcast regulation until about 1970. The surge of interest at the beginning of the decade can be attributed to several developments. The first was a presidential task force on communications policy headed by Eugene Rostow in the last years of the Johnson administration. "Rather than taking the place of competition in markets having pronounced natural monopoly features, regulation in the communications industry . . . has at times acted as a constraint on competition even in markets which do not have such features," the task force concluded.[10]

A second source of reform ideas was the White House Office of Telecommunications Policy, established in 1970 by Clay T. Whitehead. Whitehead was responsible for the "open skies" policy, which led the FCC to institute a competitive domestic satellite industry. Left to its own devices, the FCC might have permitted the old Bell System to monopolize domestic communication satellites, or at least placed substantial barriers in the path of new entrants. In addition, Whitehead favored reform of cable television and a number of initiatives regarding telephone regulation, including the notion that local and long-distance telephone service should be operated independently.

The FCC during the Nixon and Ford administrations showed little interest in regulatory reform in broadcasting, but by the time of the Carter administration, not only the FCC but congressional leaders supported change. The FCC chairman, Charles Ferris, was a vigorous and effective reformer. He abolished many restrictions on cable television and initiated a Network Inquiry Special Staff to review regulation of the broadcast networks. His successor, Mark Fowler, a Reagan appointee, became an even more implacable reformer. Fowler attacked the public trustee rationale for regulating broadcasting, suggested relaxing ownership restrictions, and endorsed auctioning off the broadcasting spectrum (Fowler and Brenner, 1982).

A third reason for reform was the interest of several foundations—Ford and Markle, in particular—in funding academic research on communications policy. The research results were disquieting. Little of the existing policy structure was found to have merit. As more and more economists and legal scholars became interested in broadcast regulation and as the flow of published research increased, the perceived legitimacy of the FCC's policies and, worse, its motives, waned. How the reduced intellectual legitimacy of the FCC's regulation of broadcasting was transformed into effective political reform has been debated by many scholars.[11] But it is clear that regulatory reform at the FCC was part of a broader, bipartisan reform movement in the mid- and late 1970s that encompassed air and surface transportation, financial services, and other regulatory regimes.

Another possible explanation of the reform movement at the FCC and perhaps at other regulatory agencies concerned the climate of the times. Since the New Deal, the intellectual framework of regulation had been established by private lawyers in Washington, D.C., who worked for the interests being regulated. Prominent among these lawyers were the founders of many of the New Deal regulatory agencies. They had invented modern regulatory institutions and administrative procedure, but their successors in office at the regulatory agencies in the 1950s and 1960s were seldom of the same caliber. Although the agencies were founded on a public interest theory of regulation, regulatory policies were largely fashioned by Washington lawyers to serve their clients' economic ends. Regulators identified too closely with the interests of those they regulated. The reconciliation of the public interest objectives of regulation with the reality of service to private interests required a logic that was vulnerable to analytic attack. The mood of the country in the late 1960s and early 1970s became less tolerant of authority and more inclined to question precedent. Stripped of respect, many regulatory institutions began to fall of their own weight.

Changes in the Industry

The most significant changes in television regulation in the 1970s affected satellites and cable. Whitehead's open skies policy, once it was accepted by the FCC, made distribution of television programs to stations and to cable systems much less expensive. Previously, television distribution had relied on costly microwave interconnections supplied by AT&T that were priced to discourage part-time and occasional users. The open skies policy attracted programmers and new networks and simultaneously encouraged UHF stations and the new cable systems. In 1976 the FCC deregulated private "receive-only" satellite antennas. Quickly these "dishes" began to dot the rural landscape. The ability to bypass cheaply the ponderous AT&T distribution system removed a significant barrier to entry into the networking business.

Also in 1977 the U.S. Court of Appeals for the District of Columbia decided that the FCC rules restricting pay programming on cable had no constitutional basis. Pay programming provided the essential impetus for cable growth in metropolitan areas with good over-the-air television services. This growth in turn created a demand for new cable networks.

Other important deregulatory initiatives followed. In 1979 most of the FCC rules restricting cable television from competing with broadcasters were removed. In 1981 the Commission eased the financial viability requirements needed to sell a station and eliminated other annual paperwork requirements.[12] Ascertainment, the practice of sounding out community needs before obtaining or renewing a license, was dropped as a requirement for radio in 1981 and for commercial television in 1984.[13] Limits on the number of broadcast stations a single company could own were eased—from 7 AM stations, 7 FM stations, and 7 television stations to as many as 12 of each type, as long as the combined television audience remained less than 25 percent of the U.S. population. A significant change by the FCC in 1988 was the abolition of the fairness doctrine, which had required the Commission to regulate broadcast content.

The FCC did not limit itself to removal of regulations restricting competition. It also promoted competition. About 4,000 stations of a new class—with lower power, so as not to interfere with existing stations—were authorized, as were new AM stations.[14] New VHF-TV stations, called "drop-ins," were also added to the list of broadcasters. (For a more detailed summary of these reforms, see Carter, Franklin, and Wright, 1986, pp. 48–49.)

Instructional television channels were converted for multichannel, multipoint distribution, or microwave transmission with a 25-mile range to fill the space of empty VHF channels on the dial.[15] The number of independent (non-network) television stations on the air grew steadily, gradually approaching the percentage of national penetration necessary to support a fourth conventional network, and meanwhile creating a healthier demand for syndicated Hollywood programs (see Figure 1.2). Barter syndication grew dramatically. The media magnate Rupert Murdoch purchased the Fox studios and a chain of television stations. In 1986 he announced the formation of a fourth broadcast network, Fox, with affiliates from Portland to Tampa. The following year Fox began offering programming one night a week, with plans to increase by one night each

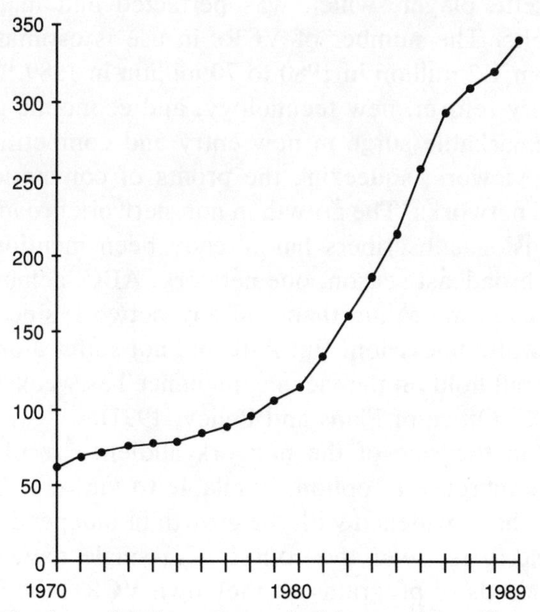

Figure 1.2 Number of independent television stations. Data from INTV (Association of Independent Television Stations), Washington, D.C.

year.[16] In the summer of 1989, Fox for the first time beat out one of the three original networks, ABC, in the weekly prime-time ratings. Early in 1990 several dramatic announcements were made concerning prospective direct broadcast satellites that would deliver signals directly to the home, bypassing local station and cable systems.

The 1977 decision by the D.C. Circuit Court of Appeals to invalidate the FCC's restrictions on pay television was a bolt from the blue.[17] Many observers had feared that free television had too much political support to permit the legalization of pay television. Yet with legalization came the rapid advance of pay services such as HBO. The revenues from pay-television subscribers are used to compete with the networks for quality programs. HBO, with only a fraction of the viewers of a broadcast network, achieves comparable revenues because viewers are willing to pay much more for programs than advertisers are willing to pay for viewers. Pay cable grew from 1.5 million subscribers in 1977 to an estimated 42 million in 1990.[18] In the late 1970s came another new development, the video cassette player, which was perfected and marketed to a willing public. The number of VCRs in use is estimated to have jumped from 1.2 million in 1980 to 70 million in 1989.[19]

Regulatory reform, new technology, and economic growth produced a remarkable surge in new entry and competition for programs and viewers, squeezing the profits of conventional broadcasters and networks. The growth in non-network broadcasters and cable television subscribers has already been mentioned. In the 1986–1987 broadcast season, one network, ABC, achieved a lower average prime-time rating than had any network since the era of black-and-white television. But ABC did not suffer alone. The networks' overall hold on the viewing audience has weakened considerably (FCC, Office of Plans and Policy, 1991).

The fall in the size of the network audience results from the tremendous increase in options available to viewers. Over-the-air alternatives have widened with the growth of independent stations. Cable subscribers, over the course of a single day, can choose among hundreds of programs. If they own VCRs, they can record one program while watching another, no longer even limited by simultaneous broadcasts. A nearby retail video outlet may offer hundreds more choices. FCC's longtime policy of protecting the pocketbooks of broadcasters restricted viewers' choices. But once the FCC changed its policy, a video smorgasbord became available for viewers.

The Notion of a Public Good

Television programs and the electromagnetic signals that carry them are public goods. This fact is important because it affects the relationship between competition and economic efficiency. There are two kinds of goods or commodities: "private" and "public." The distinction has nothing to do with who provides the good. A private good can be produced by the government. A pure private good or commodity is one whose cost of production is related to the number of people who consume it. A pure private good, if consumed by one person, is no longer available for someone else. A slice of bread is an example of a pure private good. A pure public good, by contrast, is one whose cost of production is independent of the number of people who consume it; more precisely, one person's consumption of such a good does not reduce the quantity available to other people. The standard example of a pure public good is national defense or national security.

Most goods have some elements of "privateness" and some elements of "publicness." Motion pictures and books, for instance, exhibit this dual property. Their content is a public good, but they are delivered to customers in the form of a private good. The economic conditions for efficient allocation of a pure public good or a pure private good differ markedly. A private, decentralized, competitive market system will not produce an adequate supply of pure public goods. Private production of a public good sometimes requires protection of the producer from entry or competition—in other words, monopoly power; it also may require "exclusion devices" or mechanisms for denying the public good to nonpaying consumers ("free riders"). An example of such a device is a toll booth on a public highway. Even a monopoly producer of a public good from which free riders can be easily excluded may need to practice price discrimination among its customers.

The reason for this should be clear: because it costs nothing extra to supply an extant public good to any added consumer, it is inefficient to exclude any consumer who values the good. A uniform price will exclude all consumers whose value for the good lies below that price. Given the need for the producer to receive at least enough revenue to cover costs (thereby ruling out the uniform price of zero), no uniform price is efficient. Efficiency requires that each consumer be charged a price that is no greater than the value that customer places on the public good. Similarly, the producer must charge each consumer no less than the producer's valuation of the good, or else the producer will not be able to produce the right quality or level of the public good.

Television is very nearly a pure public good. The cost of producing a television *program* is independent of the number of people who will eventually see it. (The production cost, however, may very well influence how many people will *want* to see it.) The program, whether on tape or film, remains available and unchanged no matter how many viewers see it.

If the program is broadcast, the broadcast itself is a public good, at least within the geographical area of the signal and, if scrambled, to those with a descrambler. The receipt of the signal does not reduce the quantity of the signal available to others. In this respect a television broadcast is unique. Most entertainment is heavily infected with "publicness" but is delivered to the consumer in or through a private good—a book, a magazine, a ringside seat, a theater chair. Television, however, has a public good as its delivery mechanism.

Summary

The heart of the video industry is the vertical chain of program producers, distributors (such as networks), and retailers (stores, systems, and stations). Their object is to obtain viewers in order to sell them to advertisers, to charge them an admission fee, or both.

The video industry is a regulated industry. The government develops and implements telecommunications policy. In the United States, as in other democratic societies, the process by which government makes and implements policy can be influenced by individual firms, acting alone or through associations. In the past, firms and industries made effective strategic use of their potential to shape government policy. Those who did not succeed in influencing policy became the victims of those who did.

The product sold in the video industry is a public good—that is, one with high fixed costs and low marginal costs. To be profitable, business strategies for selling public goods must repeatedly exploit each product. Competitive advantage lies in reaching the largest audience for each product and in exposing the product in as many different markets as possible.

2 / The Supply of Programming

A complex industry supplies the programs carried by television stations, cable and broadcast networks, video rental stores, and the other program delivery services described in Chapter 1. Who are the most important producers? Do a few firms dominate? Is entry easy? Are there persistent pockets of profits that are not competed away? Is competition strong enough to force suppliers to produce and price efficiently? Economists generally address these issues by examining an industry's structure and the factors that facilitate or impede entry, and we discuss program supply from this perspective in the last section of this chapter. But first we explore the strategically more interesting phenomenon of *windowing*—releasing a program in different distribution channels at different times. The practice of windowing underlies much of the intricate web of competitive interactions in the video industry. We will also look at foreign markets for U.S. programs. In one sense, foreign markets comprise a different sequence of windows, separated geographically from those in the United States. But foreign markets deserve a closer examination, because their financial importance to the U.S. television industry is quickly increasing and because cooperative ventures between U.S. and foreign production companies have rapidly grown in number. For both reasons, foreign markets will exert increasing influence on the programs offered viewers in the United States in the future.

Windowing

Windowing is one of the ways in which the television industry exploits the public-good characteristics of programs. Because production costs are fixed, cost per viewer declines with audience size.

In a competitive market, the benefits of spreading fixed costs over large audiences favor program delivery services that are able to reach large audiences. Such a goal may focus attention on the geographic expanse of a distribution system and the populations it encompasses, but there is also a temporal dimension to the construction of audiences. When a program is re-aired or made available in another format, there are always some viewers who have never seen it before and others who will watch it over and over again. A program's audience, therefore, can continue to accumulate over time with repeated broadcasts.

At one level, windowing is a natural outgrowth of the public-good character of the product, just as is the tendency for the market to favor distribution systems with a broad geographic reach. Windowing sequences are not randomly determined, however. On a per viewer basis, programs sell at prices that vary widely across distribution media and among different windows using the same medium. Windowing can be a carefully crafted strategy for maximizing the profits generated by a video product.

When a firm sells the same product at different prices to different buyers, the seller is discriminating in price among buyers. Windowing is best understood as a form of price discrimination. The producers of television programs (including feature films that are released on video cassette and shown on broadcast and cable television channels following theatrical release) sell the same public good to different program services at different prices.

Price discrimination is a common feature of market economies. Automobile manufacturers offer options that raise the price by more than the cost of adding them to the car. Airlines offer first-class and coach compartments that segregate air travelers according to their willingness-to-pay. Movie theaters offer tickets that are less expensive for children under twelve. Senior citizens receive discounts on a variety of services.

As long as buyers differ in how much they are willing to pay for a product, sellers can *theoretically* make more money by selling their product at prices that reflect its value to individual buyers than by selling at a single price. A single price generates less than maximum profits because there are generally some buyers who are paying less than they would be willing to pay and other potential buyers who are willing to pay more than the seller's marginal cost but who are excluded from the market because the price is too high.

Consider the example presented in Table 2.1. Suppose you own the rights to an exercise video that you can duplicate and distribute by cassette at a cost of two dollars per sale. There are four potential customers for your cassette. Each would like a single copy and is willing to pay the amount shown in the table. If you were forced to set a single price, the best you could do would be to price the cassette at $7 and earn $10 in profits on sales of $14 to A and B. A higher price would exclude B and limit profits to at most $8. Revenue could be increased from $14 to $15 by lowering the price to $5 since C would then buy, but after subtracting the extra $2 of production and distribution costs, profits would be reduced to $9. Profits would be lower still, $8, if the price were dropped to $4 to sell to D.

Suppose you knew that A and B patronized a video store on the north side of town and C and D shopped at a video outlet on the south side of town. You could then increase your profits through price discrimination by selling the cassette for $7 in the north-side store and selling it for $4 in the south-side store. Profits would increase to $14 on total sales of $22. This assumes, of course, that north-side buyers are unwilling to travel to the south side to save the $3 difference and that south-side customers don't find it profitable to buy extra copies and take them to the north side of town to sell at a price of less than $7. Sales and profits both could be increased by an additional $4 if you could find a mechanism for charging each buyer the full amount he or she is willing to pay for the cassette.

For price discrimination to be an effective means of increasing

Table 2.1 Four customers' willingness-to-pay for one video

	Customers			
	A	B	C	D
Maximum price	$10	$7	$5	$4
Profit per unit sold	$8	$5	$3	$2
Units sold at single price	1	2	3	4
Profits with single price	$8	$10	$9	$8

profits, the seller must be able to (1) distinguish among buyers with different demands and (2) prevent resale among buyers. If the seller cannot distinguish buyers willing to pay a high price from those willing to pay less, all customers will represent themselves as buyers with a low willingness-to-pay. Even if the seller can tell one type of buyer from another, if low-price buyers resell the product to those willing to pay a higher price, the original seller will get only the lower price. Rarely, if ever, are sellers able to extract the maximum every buyer is willing to pay. Rather, price discrimination schemes tend to distinguish among broad classes of buyers, setting a common price for each buyer class even though there may be considerable variation among buyers within a class.

Distributors of television programs and motion pictures discriminate among audiences by releasing their products at different times (windows) and in different distribution channels. At one time motion pictures were released in a series of "runs," starting with "first-run" theaters in big cities and working down to neighborhood theaters. Today the primary domestic distribution channels are cinemas, video cassettes and video disks, pay-per-view cable, pay cable, basic cable, broadcast network television, and syndication to local broadcast stations. The same channels exist in most foreign markets. In the future it may be necessary to add to this list windows for direct broadcast satellite (DBS) and multipoint, multichannel distribution service (MMDS) or other new distribution technologies. Staggering releases forces buyers to sort themselves out according to how much they are willing to pay for the film or program close to its original release date. Trade press estimates of the typical release sequence for a successful American feature film in the late 1980s are shown in Table 2.2. It does not include some of the finer details of windowing, such as a brief pay-per-view cable window that begins about a month after the home video window opens. The first American cable window is generally for release to pay channels such as HBO, and in the second cable window the film will be carried by basic cable services. Successful network television series follow a similar though shorter sequence of releases, beginning with the broadcast networks window.

Release windows have changed dramatically since the early 1970s, and they continue to evolve. The basic and pay cable windows were, at best, afterthoughts for distributors until the mid-

Table 2.2 Release windows for U.S. theatrical films in the late 1980s

Window	Months from initial release
Theaters	0–4+
Overseas theaters	4–18+
Home video	6–30+
Overseas home video	9–24+
First cable run	12–36+
Broadcast networks	36–60
Overseas broadcasters	48–60
Second cable run	66–72+
Syndication to local television stations	72+

Source: The Economist: A Survey of the Entertainment Industry, December 23, 1989, p. 5.

1970s. The cassette window developed early in the 1980s, for all practical purposes, and the pay-per-view window is even more recent. The windows for the newest distribution technologies have opened as their audiences and revenues have grown. More interesting, however, than the windowing strategies employed at one time are the principles that govern their design.

Windowing strategies are designed to maximize the profits a program can realize in all distribution channels. The strategies take account of six important factors: (1) differences in the per viewer price earned in the different distribution channels; (2) differences in channels' incremental audiences, by which we mean differences in the number of new viewers they contribute to a program's total audience; (3) the interest rate as a measure of the opportunity cost of money; (4) the extent to which viewers exposed to a program through one channel are eliminated from its potential audience in other channels; (5) differences among channels in their vulnerability to unauthorized copying; and (6) the rate at which viewer interest in a program declines following its initial release.[1]

We will explore the influence of the first three factors on windowing strategies by comparing profits from different distribution strategies in a simplified example with two distribution channels. For simplicity, assume that theaters and video cassettes are the only two distribution channels and that the potential audience for a feature film is 1,000 viewers, 700 of whom have VCRs. The remaining 300 viewers can be reached only through the cinema. The owner of a feature film must decide how to use the two distribution channels to maximize profits. The film owner knows that a film like this one attracts 15 percent of its viewers during its theatrical run, and that of these 150 viewers, 100 are drawn during the first period following the movie's release and 50 in the subsequent period, *if the cassette release is delayed to the start of the third period.* (We are following economic convention in using "period" to refer to an arbitrary, but standardized, unit of time such as a month or a year.) Two periods are sufficient to exhaust the potential cinema audience, regardless of when the film is released on cassette.

In addition, the film owner keeps in mind three truths, learned from experience. First, until a film is released on cassette, owning a VCR does not change a viewer's willingness-to-pay to see the film in a theater, and the theatrical release draws equal fractions of viewers who own VCRs and viewers who do not.[2] Second, because the video rents for substantially less than the price of a movie, viewers with VCRs will not watch the film at the cinema once it is out on cassette. Third, all viewers who are going to rent the video will do so during the period immediately following its cassette release.

The film owner has three practical distribution options: (1) give the film a two-period exclusive cinema run and release it on cassette in the third period; (2) shorten the exclusive cinema run to one period and release it on cassette in the second period; (3) release the film to theaters and on cassette in the first period. These are the three options referred to in Tables 2.3 to 2.6. The cinema run could begin after the film is released on cassette, but this would needlessly delay the receipt of cinema profits (reducing the present value of total earnings) and contribute nothing to gross profits. Because profits from the film will accrue over two or three periods, depending on the release strategy employed, the film owner must use some criterion for comparing and aggregating the profits generated in different periods. We assume that the owner will time the releases of the film in the two channels to maximize the present value of the multiperiod stream of profits the film generates.[3] This is the standard assumption of economists in multiperiod profit analyses.

Table 2.3 gives nominal (undiscounted) values and present values

Table 2.3 Profitability of three release options

	Option 1	Option 2	Option 3
Period 1			
Cinema viewers	100	100	30
Cassette viewers	0	0	420
Cinema profits	$100.00	$100.00	$30.00
Cassette profits	0	0	$210.00
Period 1 profits			
Nominal	$100.00	$100.00	$240.00
Present value[a]	$100.00	$100.00	$240.00
Period 2			
Cinema viewers	50	15	15
Cassette viewers	0	350	0
Cinema profits	$50.00	$15.00	$15.00
Cassette profits	0	$175.00	0
Period 2 profits			
Nominal	$50.00	$190.00	$15.00
Present value[b]	$45.45	$172.73	$13.64
Period 3			
Cinema viewers	0	0	0
Cassette viewers	315	0	0
Cinema profits	0	0	0
Cassette profits	$157.50	0	0
Period 3 profits			
Nominal	$157.50	0	0
Present value[c]	$130.17	0	0
Total Profits[d]			
Nominal	$307.50	$290.00	$255.00
Present value	$275.62	$272.73	$253.64

a. The present value of a dollar in period 1 is one dollar.
b. For a per period interest rate of i, the present value of a dollar in period 2 is $\$1/(1 + i)$.
c. The present value of a dollar in period 3 is $\$1/(1 + i)^2$.
d. Totals may differ slightly from the sum of column entries due to rounding.

for per period profits and total profits for each of the three distribution options when the per period interest rate is 10 percent, 60 percent of VCR owners who have not seen the film in the theater will rent it on cassette, and the film owner earns one dollar for each viewer who sees the film in a theater and fifty cents for each viewer who sees the film on cassette. Cinema attendance is highest with option 1, because the cassette release is delayed until the demand for theatrical exhibitions is exhausted. Because the film owner earns more on a viewer who sees the film in a theater than on a viewer who watches the film on cassette, option 1 produces the largest aggregate profits (undiscounted). The cost of maximizing profits on the theatrical run is the reduced present value of earnings on cassette sales that are deferred to the third period. Cassette profits can be advanced from the third period to the second period if the owner is willing to accept lower profits from the 35 viewers who would have seen the film in the cinema but now will choose the less profitable option of watching it on cassette if the cassette is released in the second period. Cassette profits can be advanced to the first period by releasing the cassette then, but at the additional cost of fifty cents forgone for each of the 70 more viewers who will rent the cassette instead of going to the theater. In this situation, the film owner will choose option 1 because this produces the largest present value of profits over the three periods.

Higher interest rates, a larger audience for the film on cassette, or an increase in the profit margin on cassette rentals relative to theater sales would make options 2 and 3 increasingly attractive relative to option 1. This is because each increases the amount by which the present value of cassette profits are reduced by delaying the cassette release to the third period. The effect of changing the interest rate on the timing of the cassette window is shown in Table 2.4. Option 1 provides the highest present value of profits as long as the interest rate is not much more than 10 percent. At the lower

Table 2.4 Effect of changes in the interest rate on the profitability of three release options (present values of three-period profits)

Interest rate	Option 1 ($)	Option 2 ($)	Option 3 ($)
.05	*290.48*	280.95	254.29
.10	*275.62*	272.73	253.64
.15	262.57	*265.22*	253.04
.20	251.04	*258.33*	253.50
.25	240.80	*252.00*	*252.00*
.30	231.66	246.15	*251.54*

Note: Calculations assume that 60 percent of VCR owners will rent the film on cassette if they have not seen it in the theater first and that the film owner realizes a fifty-cent profit for each viewer who sees the film on cassette. Figures in italic identify the profit-maximizing option for a given interest rate.

rates of interest, it pays to delay the cassette release to the third period to maximize the higher per viewer profit from cinema sales. However, for somewhat higher rates of interest (up to 25 percent), the second-period cassette window of option 2 produces the highest present value for profits. For very high rates of interest (above 25 percent), the present value for profits is maximized with option 3. With interest rates at this level, the present value of delayed cassette profits erodes so fast that the present value of total three-period profits declines if the cassette release is postponed beyond the first period.

Table 2.5 illustrates the effect of changes in the fraction of VCR owners who will watch the film on cassette if they have not seen it in the theater. Recall that the total cinema audience will top out at 15 percent of all viewers if the cassette release is delayed to the third period. We have assumed that all viewers with VCRs who are willing to watch the film in the theater will instead rent the cassette if it is released on cassette early enough. This means that the cassette release makes a net contribution to the total audience only if more than 15 percent of viewers with VCRs will watch it on cassette. The larger the net VCR contribution to the film's audience—that is, the larger the amount by which the percentage of viewers with VCRs willing to rent the cassette exceeds 15 percent—the

Table 2.5 Effect of changes in the percentage of VCR owners willing to rent the cassette on the profitability of three release options (present values of three-period profits)

Percentage of VCR owners willing to rent	Option 1 ($)	Option 2 ($)	Option 3 ($)
15	*143.48*	128.26	95.54
30	*183.18*	173.91	148.04
45	*222.87*	219.57	200.54
60	262.57	*265.22*	253.04
75	302.27	*310.87*	305.54
90	249.97	356.52	*358.04*

Note: Calculations assume an interest rate of 15 percent and that the film owner realizes a fifty-cent profit for each viewer who sees the film on cassette. Figures in italic identify the profit-maximizing option for a given percentage of viewers willing to rent the cassette.

more costly the loss in present value associated with delaying the cassette release. Therefore, a larger potential cassette audience will encourage an earlier cassette release. Option 1 maximizes the present value of profits when the percentage of VCR owners willing to rent the cassette is 15, 30, or 45 percent. Option 2 is best when this figure is 60 percent or 75 percent. And option 3 maximizes the present value of profits if the percentage of VCR owners who will watch the film is 90 percent or more.

The cassette's incremental audience is also affected by the fraction of viewers who own VCRs. We have assumed a VCR penetration rate of 70 percent for all of the calculations presented here. However, as long as more than 15 percent of viewers with VCRs are willing to watch the film on cassette, an increase in the fraction of viewers with VCRs has an effect on the optimal release strategy similar to the effect of increasing the fraction of VCR owners willing to watch the film on cassette: namely, that the optimal date for releasing the film on cassette moves closer to the theatrical release date.

Finally, we see in Table 2.6 that the larger the profit per cassette viewer relative to the film owner's profit on a cinema ticket, the more profitable it is to advance the date of the cassette release. This is because aggregate cassette profits and the cost in diminished present value from delaying their receipt grow if the profit margin on cassette rentals increases, while the opportunity cost of lost

Table 2.6 Effect of changes in the cassette rental price on the profitability of three release options (present values of three-period profits)

Cassette rental price ($)	Option 1 ($)	Option 2 ($)	Option 3 ($)
.3	*214.93*	204.35	169.04
.4	*238.75*	234.78	211.04
.5	262.57	*265.22*	253.04
.6	286.39	*295.65*	295.04
.7	310.21	326.09	*337.04*
.8	334.03	356.52	*379.04*

Note: Calculations assume an interest rate of 15 percent and that 60 percent of VCR owners will rent the film on cassette if they have not seen it in the theater first. Figures in italic identify the profit-maximizing option for a given cassette rental price.

profits on theater sales displaced by cassette rentals falls as the two profit margins converge.

Even in this simple two-media example, the balance between present value considerations and profit-per-viewer considerations is delicate. Because viewers are willing to watch the film in only one of the two distribution media, the film's owner must choose the distribution strategy that optimizes the audience trade-off between the two media. The optimal strategy will vary depending on the fraction of the viewing population with VCRs, the rate of interest, relative profit margins in the two media, and the strength of viewers' preferences to see the film in the cinema rather than on cassette.

Although the film owner's task here is highly simplified, this example reveals the types of trade-offs among channels that must be considered in the design of windowing strategies in the real world. A viewer who has seen a film on cassette is unlikely to pay to see it in the cinema, and a viewer who has seen a film on cable or free television will be less inclined to pay $2 to $4 to see it on cassette. Of course, these relationships hold in reverse. A viewer who has seen a film at the cinema will be less inclined to rent it on cassette, and those who watch the cassette probably will not be in the audience for subsequent cable or broadcast showings. Similar trade-offs exist for programs introduced in one of the broadcast or cable windows. The timing and length of release windows in various media will reflect these audience trade-offs as well as interest rate and audience size considerations.

Waterman (1987) estimated a producer/distributor's per viewer profit margins in various release windows (see Table 2.7).[4] The ranking by profit margins corresponds closely to the timing of releases into these windows illustrated in Table 2.2, an indication of the importance of profit margin in the formation of distribution strategies.[5] Release strategies have changed over time. Before cable, major motion pictures went to the broadcast networks immediately following their initial cinema runs. As cable and the VCR developed as distribution channels and proved to have higher profit margins, they were given new windows before network television in the release sequence. Changes in the timing and ordering of releases into different media during the 1980s reflected the importance of the incremental audiences contributed by the newer media.

Table 2.7 Prices and net revenues from theatrical feature film distribution (1984)

Media	Effective retail price per viewer	Net producer/ distributor revenue per viewer (excluding advertising)
Theaters	$3.50	$1.00
Prerecorded video cassettes		
Sales[a]	6.25–12.50	1.87–3.75
Rentals[b]	1.33	.28–.42
Pay-per-view cable TV[c]		
Monthly subscription	1.00–1.25	.50–.62
Pay cable TV/STV[d]	1.00	.20
Network broadcast TV[e]		.05
Syndicated broadcast TV[f]		.01

Assumptions:
a. $50 retail price, 4 to 8 viewers per purchased tape.
b. $4 retail price, 3 viewers per rental transaction; 12 to 18 rental transactions per tape.
c. $3–$4 charge per household, 3 viewers per household.
d. $10 monthly charge, 2 viewers each for 5 new theatrical movies per month.
e. Three runs, 15 average rating per run, 2 viewers per household for each run.
f. Ten runs, 4 average rating per run, 2 viewers per household for each run.

Source: Reprinted with permission from David Waterman, "Electronic Media and the Economics of the First Sale Doctrine," table 3, in 1987 *Entertainment, Publishing, and the Arts Handbook,* ed. R. Thorne and J. D. Viera, © Clark Boardman Company, Ltd., New York, N.Y.

The cassette window advanced closer to the date of theatrical release as the fraction of television households owning VCRs mushroomed from near zero at the beginning of the decade to approximately two-thirds at its close. As cable penetration increased, the basic cable window, which had always followed broadcast syndication, began to receive packages of feature films that previously would have gone to broadcasters before being released to cable services. Cable networks also stepped in to provide the off-network premieres of certain dramatic series that in earlier years would have been syndicated to independent stations following their runs on the major broadcast networks.[6]

Three factors that affect windowing strategies were not covered in this example—the willingness of viewers to watch a program more than once, copying, and the time sensitivity of the demand for a program. The trade-off of viewers among channels is less severe if there are viewers who will watch a program repeatedly as it appears in different windows. Repeat viewing does occur, although we know from the continuous supply of new programs and the audiences they receive that most viewers prefer new programs.[7] To the extent that a program supplier can profit by selling a program more than once to the same viewers, the supplier has an incentive to prolong the earlier release windows in which per viewer profits are higher.

Unauthorized copying is a serious threat to the financial well-being of program producers. Copying short-circuits the windowing process and denies the producer the full benefits of price discrimination. To postpone the anticipated harmful effect of copying on revenues, producers delay releases into the distribution channels that are most vulnerable to copying.[8] Copying of video products is less of a problem in the United States than it is in some other countries.[9] Motion picture producers have in some cases refused to license their films for release on cassette in countries where pirated cassettes dominate prerecorded video sales.

For many live programs, such as athletic events and rock concerts, the demand for subsequent showings is but a tiny fraction of the original demand. Interest in the program drops dramatically after the event—so much so that many events are never rebroadcast. For this reason, the period over which the program is released into different media may be compressed dramatically. Second showings of sports events tend to occur soon after the actual date of the event. Major boxing matches are released on video shortly after the fight, and sports channels often repeat basketball or football games in their home markets late at night on the day of the game for viewers who could not watch the contest when it was televised live. The more time sensitive the demand for a particular program, the faster it must be cycled through the various release windows.

Competition among Producers

Windowing is an important element in competition among program producers, who compete with each other in at least four ways. First, they compete to sell their programs to program services. Second, they compete for the talent required to make programs.

Third, they compete indirectly for audiences, because their success depends on the appeal of the schedule of programs that they and the program delivery service jointly create. Fourth, they compete with each other over time, because program services consider previously exhibited programs as substitutes for new programs. Producer competition would exist without windowing, but windowing strengthens its intertemporal dimension and thereby affects its other aspects as well.

Intertemporal competition may be thought of as competition among the producers of programs of different vintages. If programmers find the cost of new programs to be too high, they may turn to programs that have been shown before (on the same or other distribution channels) that can be acquired for less.[10] The demand for new programs, both now and in the future, is constrained by the supply and characteristics of old programs still available for broadcast. Windowing is not required for intertemporal competition. Almost all program services repeat the programs in their schedules to some degree (Wildman and Lee, 1989). For example, broadcast network series rebroadcast as summer reruns are programs that are broadcast twice during the broadcast network window. However, windowing greatly facilitates the reuse of programs by providing additional outlets on which they may be shown.

How important is the supply of old programs as a constraint on prices paid for new programs? In theory, the effect could be quite large, although in practice, it probably cannot be quantified. Product durability may significantly limit the ability of firms to price above cost, as when the durability of old cars limits the prices auto manufacturers can charge for new cars because consumers respond to higher prices by driving their old cars longer. In the extreme case of an infinitely durable product, even a monopolist may be forced to sell at a competitive price.[11] Programs recorded on film and tape may, with care, be preserved indefinitely. Most television programs (except a few classics) lose audience with repeated exposure, however, and even without exposure their content becomes dated, which eases the pressure of old programs on the prices paid for new programs.

In addition to strengthening the intertemporal character of competition, windowing makes the competition more direct among firms producing programs for different distribution channels.

Because programs can be repeated in other channels, profit-maximizing producers must consider demand and supply conditions in every window in which their programs are likely to be shown. If the industry supplying programs is competitive, the prospects for ancillary earnings in all downstream windows need to be considered in program budgets and design. Each window makes a necessary contribution toward covering program costs.

In effect, this means that producers of programs that may be shown in several distribution channels are competing in each of those channels. Changed competitive conditions in one window are almost always reflected in programs released initially in other windows, as producers make strategic adjustments. Film producers thus responded to the growth of television as an aftermarket for motion pictures by placing important actors closer to the center of the picture to fit the narrower dimensions of the television screen. When the opening of cassette and cable windows for motion pictures reduced the value of theatrical films to broadcasters, the broadcast networks responded with more "made-for-television" motion pictures. More recently, improved syndication demand for half-hour off-network sitcoms and declining demand for off-network hour-long dramatic series have contributed to a rebirth of the situation comedy on the networks.[12] (Off-network programs are programs distributed by syndicators that have been aired during prime time by one of the major broadcast networks.)

Much of the popular wisdom concerning bargaining relationships among producers and program services fails to appreciate the intertemporal and interdistribution-channel nature of competition. For example, a program's earnings in the windows that follow its initial release are commonly viewed as a windfall to either its producer or the program service that carried it first. Thus producers and ministers of culture in other countries frequently complain that American programs and films cover production costs in their domestic releases, which makes it possible to sell them at "predatory" prices in foreign markets.[13] Similarly, the producers of programs for the major broadcast networks often claim that the networks are in the enviable position of being able to commission programs at substantially less than full cost, which forces producers to look to off-network syndication to make up their deficits on the network run. Both arguments fail to recognize that spreading production costs across different release windows is a natural, indeed inevitable, outcome of windowing in a competitive market for television programs.

It is true that windfalls are realized when *unanticipated* advances in program delivery technology open new distribution channels and renew demand for programs that have already been produced. MGM management could not have foreseen that "Gone with the Wind" would be a major attraction on cable television when they produced it in 1939. And who would have anticipated the continuing interest in the early 1950s episodes of "The Honeymooners" and "I Love Lucy"? Unanticipated new delivery technologies and the new windows they open disrupt established practices and contractual relationships. Well-positioned players and those first to recognize the implications of new technologies may profit enormously as the industry adjusts. However, once the profit potential of a new distribution channel and the new windows it opens are realized, producers and programmers quickly build the earnings from new windows into their production budgets and their negotiating positions. If program production is competitive, revenues generated will just cover costs incurred on average when the market is in equilibrium. Therefore, once a new equilibrium reflecting the potentials of new windows is achieved, expected revenues (measured in terms of their present values) *from all windows* will again just cover costs for new programs. Comparing production costs with revenues generated in a single window will appear to show that producers are operating at a loss. In general, one would expect that as new windows open the percentage of a program's production costs covered by the older windows would decline.

The effects of multiple windows on program budgets and the bargains struck between producers and program services are illustrated in Figures 2.1 and 2.2. Figure 2.1 depicts a situation in which there is only one distribution channel and a single window for a program on that channel. The size of a program's production budget is measured on the horizontal axis. Program revenue and total cost, the sum of production and distribution costs, are measured on the vertical axis. Overhead costs of program services are included as part of distribution costs, but they could be separated without affecting our conclusions. Distribution costs are assumed to be D. The horizontal line at D shows that distribution costs are indepen-

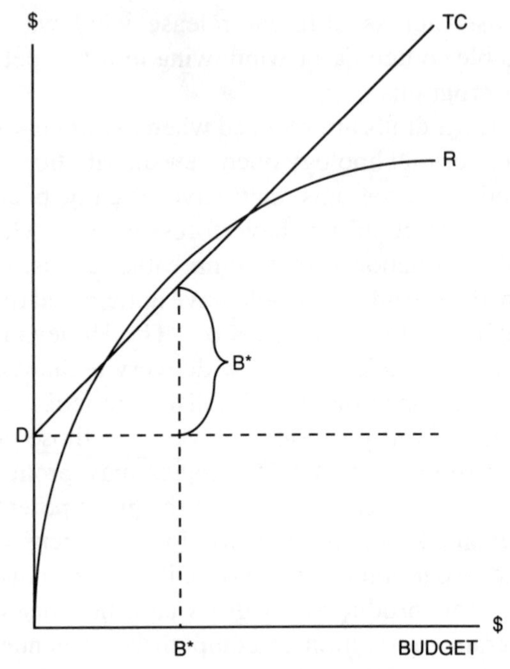

Figure 2.1 Single distribution channel, single window

dent of the size of the production budget.[14] Total costs (TC) start at D on the vertical axis (where production costs are zero) and rise with a slope of one (at a 45-degree angle). Every dollar increase in the production budget is reflected in a dollar increase in TC.

Expected revenue is represented by the curved line R. Revenue rises with the size of the budget, because additional production expenditures usually increase a program's appeal to viewers. A larger budget makes it possible to hire better writers, directors, and camera crews; to shoot from more camera angles and do more editing; to use more sophisticated special effects; to hire actors with greater name recognition and presumably the ability to deliver bigger audiences; and so on. Therefore, we would expect that larger budget productions will have larger audiences and, for pay media, audiences that will pay more to see them. Although not all expensive films and programs generate large audiences and revenues, on average additional production dollars make a positive contribution to audience and revenue. Actual revenues may be greater or less than those suggested by R. Both expensive flops and low-budget hits do occur, but R represents the revenue producers and program services expect production dollars to generate.

Although R always rises with the size of the production budget, the rate of increase (or slope) becomes smaller and smaller as the budget grows. (Another way of describing the general shape of R is to say that it is concave downward.) The expected revenue returned by an additional dollar invested in the production budget declines as the budget gets larger. For example, increasing the budget by enough to cast a major star instead of an unknown in a lead role may contribute significantly to a program's expected audience. Serious fans may watch to see the star if nothing else. A second star probably will add less to expected audience and revenue. By the time a program is "star-studded," an additional luminary probably contributes considerably more to cost than to revenue. The same logic applies to the other inputs used to produce programs.

To be produced, a program must be expected to generate revenues that will at least cover the costs of production and distribution. In terms of Figure 2.1, this means that at some point R must at least equal TC. For the situation depicted, R actually rises above TC over part of its range. Profits, measured as the vertical distance between R and TC, are positive in this region. The production budget with the largest distance between R and TC is the budget that maximizes expected profits. Expected profits rise with increases in the production budget as long as R is rising faster than TC (or the slope of R is greater than the slope of TC). As long as this relationship holds it pays to increase the production budget. Profits fall with increases in the production budget when TC rises faster than R. The profit-maximizing budget, B*, is therefore the budget at which the slopes of R and TC are equal. The budget that maximizes expected profits is determined by increasing the budget as long as an extra dollar in production expenditures generates more than a dollar increase in expected revenue.

The allocation of a program's profits between its producer and a program distribution service such as a network depends on the relative bargaining strengths of the two parties. If program production is highly competitive, competition among producers seeking to supply programs to the distribution service will force them to sell their programs at cost. Our analysis of competition among program suppliers in the preceding section suggests that program supply is very competitive. For the remainder of this section we will therefore assume that, on average, producers earn revenues sufficient to cover their costs (including the return on their investments required to keep them in business), but no more. In terms of Figure 2.1, this means that program distribution service profits equal the vertical distance between R and TC.

This analysis is extended to a program exhibited in two windows in Figure 2.2. R_1 is expected revenue from the first window, and R_2 is expected revenue from the second window. Combined revenue, R_c, is the sum of R_1 and R_2. We have assumed that revenue from the first window is greater than revenue from the second window, a relationship similar to that between the revenues generated by the broadcast networks' prime-time programs during their network runs and subsequent syndication revenues. However, this

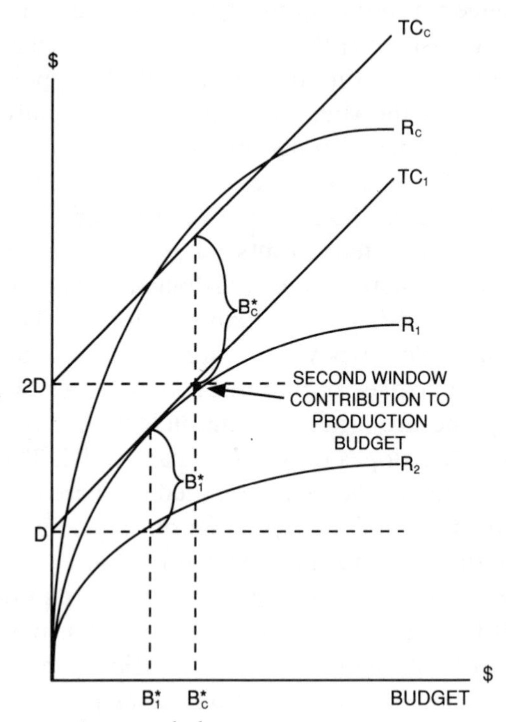

Figure 2.2 Program in two windows

ordering could be reversed without affecting the character of this analysis.

For convenience, distribution costs are assumed to be the same for both windows, so that total distribution costs are 2D if the program is distributed in both windows. TC_1 is the cost of producing and distributing the program if it is released only in the first window. TC_c is combined production and distribution costs if it is released in both windows. Not all programs are distributed in multiple windows. Certain types of programs, variety shows for example, traditionally have not done well enough in second releases to cover distribution costs. This analysis applies to the types of programs for which windowing is profitable. For the program and situation depicted in Figure 2.2, the program could cover its costs in the first window alone, but profits are greater if it is rereleased in the second window also. A budget of B_1^* maximizes profits ($R_1 - TC_1$) if the program is released in the first window only. B_c^* maximizes profits ($R_c - TC_c$) if the program is released in both windows. As with the single-window example, B_1^* and B_c^* correspond to the points on R_1 and R_c, respectively, where the slopes equal one. B_c^* is larger than B_1^* because, since $R_c = R_1 + R_2$, the slope of R_c always exceeds the slope of R_1 by the slope of R_2. Therefore, the production budget must be increased beyond B_1^* before the slope of R_c declines to one. The larger production budget is the market's response to the potential for profits in a second window. Note that R_1 is no longer sufficient to cover production costs and the costs of first-window distribution for a budget of B_c^*. The difference—the distance between TC_1 and R_1—must be made up from second-window revenues.

Changes in the revenue-budget relationships for the two windows affect profit-maximizing budgets and the implicit allocation of production costs between the first-window and second-window program services. This is illustrated in Figure 2.3 for an increase in expected second-window revenue from R_2 to R_2'. The effect is to increase the profit-maximizing production budget to B_c' and to increase the second window's contribution to production costs. Waterman and Grant (1989) argue that the development of cable as an aftermarket for motion pictures had this type of effect on the budgets for films produced by the major Hollywood studios, which increased by 122 percent (in constant dollars) from 1976 to 1985.

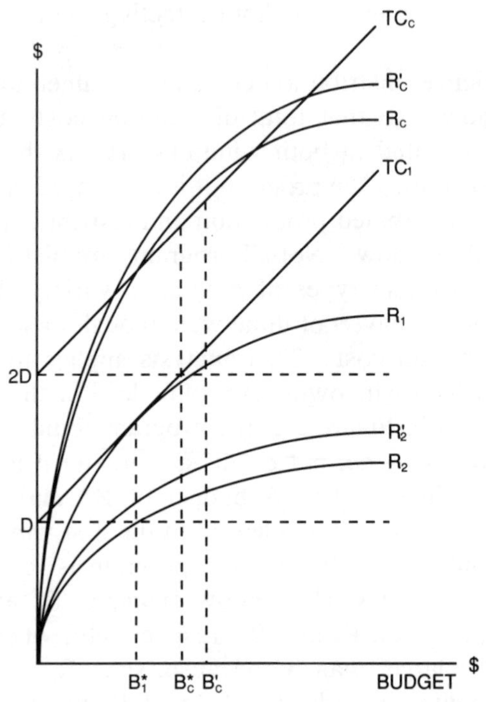

Figure 2.3 Change due to increase in expected second-window revenue

A decrease in expected second-window revenue would have the opposite effect.

The program distribution service earns positive profits in Figure 2.1, as does the first-window distribution service in Figures 2.2 and 2.3. This condition, however, cannot be assumed to hold generally. If program production is competitive and programs are sold at cost, program distribution profits are determined by the intensity of competition among distribution services. If the market for program distribution services is competitive, then such services also will just break even (see Figure 2.4). In this case, the first-window program service would be operating at a loss if it paid all of the production costs. Therefore, payments from the second-window program service must cover a portion of production costs if the program is to be produced.

The FCC's financial interest rule, before it was modified in 1991 (see Chapter 5), prohibited ABC, CBS, and NBC from owning financial interests in the domestic and foreign off-network syndication earnings of their prime-time programs. These rights are retained by producers. According to our model of budget determination and cost allocation among windows, competition among producers should result in programs' being offered to networks for less than production cost if these programs have value in off-network syndication. Average production budgets, network license fees, and producer deficits (the production costs not covered by network license fees) for programs scheduled by the three major broadcast networks at the start of the 1988–1989 and 1989–1990 television seasons are reported in Table 2.8. (Fox programs are not reflected in these averages because the Fox network was not subject to the financial interest rule during this period, and will not be until it broadcasts more than 15 prime-time hours of programming each week.[15] Programs that the networks produced for themselves, such as ABC's "Moonlighting," are also excluded.) The deficits on situ-

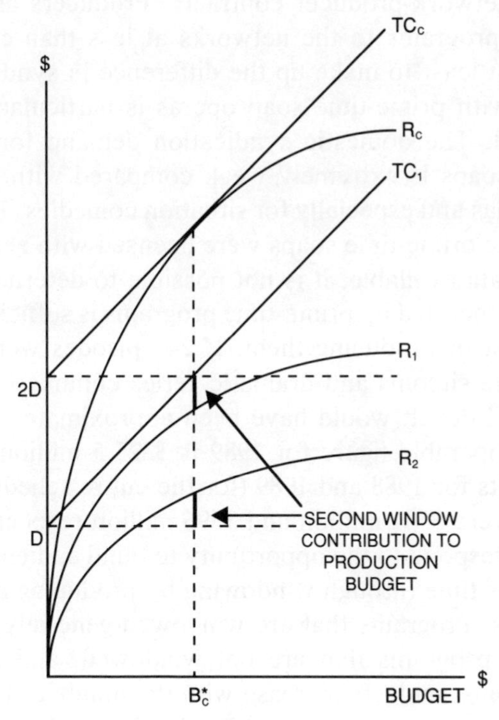

Figure 2.4 Competitive market for programming, effect on program services

Table 2.8 Producer and network share of production costs for fall prime-time series on ABC, CBS, and NBC

Type of series	Number of programs		Average budget ($000)		Average network license fee ($000)		Average producer deficit ($000)		Deficit as % of budget	
	1988	1989	1988	1989	1988	1989	1988	1989	1988	1989
Situation comedy	31	36	560	571	479	483	81	88	14.5	15.4
Hour drama	27	25	1,060	1,187	822	938	244	249	22.9	21.0
Prime-time soaps	4	3	1,388	1,400	1,388	1,400	0	0	0	0

Source: "Deficits Pending," *Channels,* October 1988, pp. 56–57; October 1989, pp. 32–33.

ation comedies and hour-long dramatic series are consistent with the hypothesis that potential off-network syndication earnings are reflected in network-producer contracts. Producers are willing to license their programs to the networks at less than cost because they expect at least to make up the difference in syndication. The comparison with prime-time soap operas is particularly revealing in this regard. The domestic syndication demand for prime-time off-network soaps is extremely weak compared with the demand for hour dramas and especially for situation comedies. For 1988 and 1989, all of the prime-time soaps were licensed with zero deficits.[16]

With the data available, it is not possible to determine whether the revenue generated by prime-time programs is sufficient to cover the entire cost of producing them. If 24 episodes were produced for each of the sitcoms and dramatic series commissioned for fall 1988, the total deficit would have been approximately $217.8 million. The comparable figure for 1989 is $225.5 million. Aggregate network profits for 1988 and 1989 (for the entire schedule, not just prime time) were $501 million and $696 million respectively.[17]

Producers respond to the opportunity to build audiences for their programs over time through windowing by producing more expensive programs. Programs that are windowed generally have larger budgets than programs that are not windowed, and among windowed programs, budgets increase with the number of windows in which the programs are released. Production budgets for motion pictures and network prime-time series clearly fit this pattern. Budgets for the programs in the network prime-time schedules averaged in the neighborhood of $1 million an hour for the 1988–1989 and 1989–1990 television seasons while production budgets for American feature films made during this period averaged from $9 million to $10 million.[18] But feature films are released in the cinema, on video cassette, and to pay-per-view and premium cable networks before they are made available to the broadcast networks.

How does windowing affect the terms on which producers compete for the talent and other production factors used to make television programs? The tendency of windowing to increase production budgets favors the producers of programs that are suitable for windowing. Producers of programs for which there is likely to be a continuing demand can, and do, spend more to produce their programs. When the demand for repeated showings of a program is weak, the producer must make do with a smaller budget and lesser talents and technical support.

Foreign Markets

Foreign sales represent about half of total sales of films and programs for the major producers (Renaud and Litman, 1985). This includes sales for cinematic releases, cassette rentals, and sales to foreign broadcasters. Cinema sales are a larger share of total sales in most foreign markets than they are in the United States. Until recently, public policy in most countries severely limited (if not prohibited) the development of commercial television. However, in percentage terms, the mix of foreign revenue has been shifting toward more television sales (Renaud and Litman, 1985). This trend accelerated dramatically during the last half of the 1980s as European countries began privatizing their television industries (Waterman, 1988; Wildman and Siwek, 1988, 1991; Dunnett, 1990).

Windowing is an important component of distribution strategies in foreign markets for the same reasons that it is important in the United States. U.S. films and programs are released in foreign markets in a series of windows that parallels, with a lag, the release sequence in the United States. Appropriate adjustments are made for differences in legal and regulatory regimes. Production budgets affect prospective foreign sales in much the same way that they affect domestic sales, after making allowance for the fact that U.S. films and programs are dubbed or subtitled in markets where English is not the native language.

In terms of Figures 2.1 through 2.4, the revenue-budget schedules for foreign markets are simply added to the schedules for domestic windows. This addition shifts the aggregate revenue-budget schedule up and out, similar to the effect of adding the revenue-budget schedule for a second U.S. window to that for the first window. The result is larger budget programs than would be produced for the American market alone. Of course, competition from foreign programs will reduce the revenue generated by U.S. programs and, to the extent that this occurs, force the production of less expensive programs. Although U.S. programs are well received in most foreign markets, American viewers have little appetite for foreign programs, especially those recorded in languages other than English. In his study of worldwide viewing patterns, Varis (1984) found that imported programs accounted for approximately one-third of broadcast hours in the foreign countries surveyed. The United States was by far the largest supplier of these imported programs. But foreign programs occupied only 1 percent to 2 percent of broadcast hours in the United States, and programs from the United Kingdom accounted for most of this.

The U.S. Office of Technology Assessment estimated the trade surplus on programs and films to be in the $1 billion range in the early 1980s, a figure that probably has grown considerably with the rapid growth in sales to newly privatized European broadcasters in recent years. The trade imbalance in these cultural products between the United States and most of the rest of the world, especially Third World countries, has been the subject of considerable debate. Some cultural theorists have argued that the imbalance reflects deliberate attempts at cultural domination (Schiller, 1969). There have also been allegations that U.S. producers compete unfairly through the distribution organizations they control (Pruvot, 1983). Economists studying the international video trade have offered a different explanation. They argue that the sizes of the markets defined by linguistic and cultural similarities influence the production budgets of programs and films produced for initial release in these markets (Hoskins and Mirus, 1988; Waterman, 1988; Wildman and Siwek, 1988, 1991). "One-way flows" in films and programs reflect such differences in market size.

Profit-maximizing production budgets will be larger if producers can build audiences and revenues over time with multiple releases in different distribution windows. Similarly, the ability to build audiences by geographically aggregating over larger populations should result in larger budgets. In addition to the large and wealthy population of the United States, U.S. producers benefit from the global English-language market, which is three to four times larger, in terms of income and population, than the Japanese and German linguistic populations, which constitute, respectively, the second- and third-largest linguistic markets for video entertainment in free market countries (Wildman and Siwek, 1988). In competing for the English-language market, U.S. producers produce much more expensive programs and films than do producers from other countries (Hoskins and Mirus, 1988; Waterman, 1988; Wildman and Siwek, 1988, 1991). All viewers prefer programs and films in their native languages to dubbed or subtitled productions. Films and programs produced in languages other than English are at a double disadvantage in competing for U.S. viewers: they have the handicap of dubbing or subtitles, and they often have inferior production values because of much smaller budgets. From an economic standpoint, therefore, it is not surprising that U.S. films and programs generate large audiences in foreign markets while foreign language productions receive only limited exposure in the United States.

The recent growth in the exports of U.S. programs has been accompanied by greater participation by U.S. producers in coproduction ventures with foreign partners. These partnerships have funded the production of high-budget miniseries and made-for-television movies for syndication in the United States and other countries. Coproduction agreements usually provide for financial participation by all partners, with each retaining exhibition rights for its own country or territory. Because the English-language market is so much larger than other linguistic markets, most of these joint venture productions have been recorded in English and dubbed or subtitled into other languages. Exceptions have been a few action-adventure films for which critical speaking scenes were shot in two or more languages.

As foreign markets continue to grow relative to the U.S. market, U.S. producers' joint ventures with foreign firms will grow in number as well. The privatization of television industries in other countries will expand the commercial potential of their markets and diminish the comparative advantage enjoyed by U.S. producers somewhat. But given the much larger English-language market, the advantage will not disappear entirely.

The Competitiveness of Program Supply

In examining the industry that supplies television programs, we concentrate on suppliers of prime-time programs to the major broadcast networks and exclude suppliers of locally originated programs. With the exception of local news, programs produced for local audiences do not fare well in competition with programs produced for wider distribution. We also consider briefly the suppliers of syndicated programs to broadcasters and the suppliers of programs to cable networks. Our focus on prime-time broadcast network programs is dictated in part by the lack of comparable data on the supply of programs during other dayparts and through other distribution channels, such as cable and independent stations. In spite of increasingly vigorous competition, the three largest broadcast networks still account for most viewing hours. The many FCC inquiries regarding the regulation of the commercial practices of these networks have focused on the provision of prime-time network programs for similar reasons.

It is no accident that programs distributed by the three largest broadcast networks have always dominated the ratings or that cable networks and independent stations began to attract substantial audiences only when they began to reach a significant fraction of U.S. television households during the 1980s. Unless there is strong local or regional variation in program appeal, programs with wide geographic distribution can be expected to attract the largest shares of audience. This is true because substantial cost savings can be realized by spreading the fixed costs of production over as large an audience as possible, and there is an incentive to produce more expensive programs when it is possible to reach a large audience. It is much less expensive to produce a single program for distribution in numerous markets than to produce separate programs for each individual market. This provides a powerful incentive to aggregate audiences by distributing programs nationally. Profit maximization also dictates more expensive programs for national audiences than for local audiences for the same reasons that windowing and international sales lead to more expensive programs. At the margin, a dollar added to the production budget contributes more to total viewership when the potential audience is large. Local stations or local cable systems alone cannot afford to produce programs that are equivalent in quality to those distributed nationally.

As discussed in Chapter 5, networks procure and schedule programs for local stations and cable operators, and they sell the advertising time within and between programs. Networks are not essential to either of these activities. Advertisers could procure programs, as many did during the early 1950s. Alternatively, program producers (or packagers) could sell time in their programs to advertisers and then sell the programs to stations, as is done in the growing barter market today. Or stations could take the initiative in contacting program suppliers and advertisers. All of these arrangements either have been used or are still used to some extent in the television industry. However, programs distributed by networks (both broadcast and cable) account for the most viewing time and the most advertising money, because the network is an efficient organizational form for arranging these transactions.

Four important advantages of networking explain the dominance of networks in nationwide program distribution. First, networking reduces transaction costs. Consider the costs of putting together a prime-time, weekday schedule. For simplicity assume that all programs are one hour long. Fifteen one-hour shows would be required to program this daypart. If each of, say, 200 network affiliates of a major broadcast network had to negotiate with the supplier of each program individually, 3,000 individual negotiating sessions would be required for all of the stations to fill their prime-time schedules. If a network bargains as an agent for the stations instead, the total number of negotiating sessions could be reduced to 215, 15 sessions for bargaining with program suppliers and 200 sessions between the network and its affiliates,[19] reducing the number of sessions required by approximately 93 percent.[20] Similar savings are realized by having networks act as intermediaries in selling commercial time to advertisers.[21]

Second, networks offer advertisers an efficient way to distribute their advertising budgets over many programs. From the advertisers' perspective, this accomplishes two desirable goals. It reduces the risk that advertising investments will be lost and critical audiences not reached if a particular program does not do well, and it also makes it easier for advertisers to achieve specific goals for the size of the audience reached and the number of times viewers are exposed to their commercials as they shift from program to program.

Third, networks can take advantage of "adjacency" effects if they select all of the programs in their schedules. (Viewers are more likely to watch a program if they have watched the previous program or intend to watch the succeeding one on the same channel.) Advertisers supplying programs would find it difficult to achieve the level of coordination required to do this effectively.

Fourth, networks reduce transmission costs by transmitting their programs simultaneously to all affiliates within a time zone. This reduces the per station costs of satellite transmission, in particular. However, this advantage is diminishing as transmission costs fall with advances in communications technology.

The industry that supplies programs to the major broadcast networks is *monopolistically competitive*. Each program is different, and therefore the criterion of perfect substitutability among products required for perfect competition is not satisfied. However, programs are sufficiently close substitutes for each other and competition among firms is sufficiently intense that, on average, firms cannot expect to earn more than competitive profits (Chamberlin, 1956).

The apparent vigor of competition among program suppliers reflects the basic structural characteristics of this industry and the nature of the product. The industry that supplies programs to the broadcast networks is unconcentrated, easy to enter, and characterized by unstable shares for the leading firms. Even if the industry were not structurally so disposed to competition, the fact that firms compete most intensely on intangible, and hard to monitor, dimensions of program quality would make collusion extremely difficult.[22]

Concentration among Network Suppliers

As markets become less concentrated, firms find it harder to develop and maintain the coordination required to hold prices above competitive levels, for three reasons. First, the coordination required to formulate and carry out a joint strategy is easier the fewer the firms involved. Second, it is easier to monitor compliance and penalize noncompliance when there are only a few competitors. Third, the less concentrated an industry, the greater the temptation to cheat on an agreement to fix prices (or product characteristics), because the proportional gain to an individual firm in sales diverted from other firms increases the smaller the potential cheater's share of sales beforehand.[23]

Tables 2.9 and 2.10 report two measures of shares of programs supplied to ABC, CBS, and NBC over a recent five-year period by program suppliers. The distinction between major suppliers and lesser, but still significant, suppliers was made in 1988 by *Broadcasting Magazine*,[24] which added Lorimar-Telepictures to the seven firms traditionally classified as major suppliers of films and television programs. What the tables refer to as supplier shares are producers' shares of the total hours of prime-time series in the fall schedules for ABC, CBS, and NBC for each year as reported in the corresponding annual issues of the *International Television and Video Almanac*. Differences in the shares reported in Tables 2.9 and 2.10 reflect different ways of classifying programs in which a major supplier participated but was not the sole producer. A program's time is divided equally among all participating producers listed in the *Almanac* for the share calculations reported in Table 2.9; the shares reported in Table 2.10 give major producers full credit for all programs in which they participate. Implicit in the Table 2.10 calculations is the assumption that the majors control the disposition of all programs in which they participate, perhaps because participating producers depend on them to finance the projects. The Table 2.10 figures should be regarded as upper-bound estimates of the major suppliers' shares of prime-time series sold to the networks.

By either measure, the supply of programs to the broadcast networks is unconcentrated. The combined shares of the major suppliers varied from the high twenties to the low thirties in Table 2.9 and from the low forties to just over fifty percent in Table 2.10. The tables also report values for the Hirschman-Herfindahl Index, an index of concentration routinely used by antitrust agencies in merger investigations to assess the likelihood that an increase in concentration caused by a merger will permit firms in an industry to exercise market power. The HHI is calculated by adding up

Table 2.9 Supplier shares as percentage of prime-time hours: proportional allocation among production partners

	1984 %	1985 %	1986 %	1987 %	1988 %
Major Suppliers					
Lorimar-Telepictures	2.8	3.0	4.9	5.3	6.3
Warner Brothers TV	3.4	2.7	3.4	3.7	3.0
Walt Disney	0.0	0.4	3.4	1.4	1.5
Columbia-Embassy	7.6	7.3	3.9	4.7	6.1
20th Century Fox	3.0	2.1	3.4	5.4	2.3
Paramount TV	1.5	2.7	1.7	1.4	3.8
Universal TV	7.2	12.1	7.2	8.4	7.2
MGM/UA	1.3	0.8	0.8	0.7	2.3
Total	26.8	31.1	28.8	29.0	32.5
Other Significant Suppliers					
Stephen J. Cannell	6.1	5.3	2.3	5.8	3.8
GTG Entertainment	0.0	0.0	0.0	0.0	3.0
MTM	5.3	5.3	3.8	2.2	1.5
New World	0.0	0.0	1.5	3.2	1.1
Orion Pictures	0.5	0.5	1.3	0.5	0.0
Republic	0.0	0.0	0.0	0.5	0.8
Aaron Spelling	7.6	4.5	3.4	2.2	0.8
Viacom Productions	0.0	0.0	1.3	2.2	0.5
Total	19.5	15.6	13.6	15.8	12.3
Other Suppliers	53.7	53.3	57.6	55.2	44.8
Hirschman-Herfindahl Index	541	573	443	398	484

Source: International Television and Video Almanac (New York: Quigley Publishing, 1985, 1986, 1987, 1988, 1989).

the squares of market shares (after multiplying each percentage share by 100) for all firms in a market. For example, Lorimar-Telepicture's contribution to the HHI in 1988 under the Table 2.10 assumptions for calculating major suppliers' shares is approximately 96, which is 9.8 squared. The antitrust agencies assume that markets with HHIs of less than 1,000 are structurally predisposed to perform competitively. By this standard, program supply to the three largest broadcast networks is extremely unconcentrated. Even with the merger of Lorimar-Telepictures and Warner, the increase in the 1988 HHI in Tables 2.9 and 2.10 would be only 38

Table 2.10 Supplier shares as percentage of prime-time hours: major partners assumed dominant

	1984 %	1985 %	1986 %	1987 %	1988 %
Major Suppliers					
Lorimar-Telepictures	4.5	6.1	8.3	9.8	9.8
Warner Brothers TV	6.1	4.5	6.1	6.1	5.3
Walt Disney	0.0	0.8	3.8	1.5	1.5
Columbia-Embassy	9.8	11.4	6.1	6.1	6.8
20th Century Fox	5.3	4.5	5.3	3.8	3.0
Paramount TV	3.0	5.3	3.8	3.0	6.1
Universal TV	12.1	17.4	10.6	10.6	8.3
MGM/UA	3.0	1.5	1.5	1.5	5.3
Total	43.8	51.5	45.5	42.4	46.1
Other Significant Suppliers					
Stephen J. Cannell	6.1	4.5	1.5	4.5	3.8
GTG Entertainment	0.0	0.0	0.0	0.0	3.0
MTM	5.3	5.3	3.8	2.3	2.3
New World	0.0	0.0	1.5	3.4	1.1
Orion Pictures	0.5	0.5	1.3	0.5	0.0
Republic	0.0	0.0	0.0	0.5	0.8
Aaron Spelling	6.8	4.5	3.4	2.3	0.8
Viacom Productions	0.0	0.0	1.3	2.3	0.8
Total	18.7	14.8	12.8	15.8	9.3
Other Suppliers	37.5	33.7	41.7	41.8	44.6
Hirschman-Herfindahl Index	719	866	615	546	627

Source: International Television and Video Almanac (New York: Quigley Publishing, 1985, 1986, 1987, 1988, 1989).

and 104, respectively, leaving the HHI still comfortably below the 1,000 level.

Ease of Entry and Share Instability

If we define entry as movement from zero to a positive share of network prime-time hours, then five of the sixteen firms in Tables 2.9 and 2.10 (including one of the major suppliers, Disney), entered between 1984 and 1988. Disney, of course, is a longtime supplier of network programs that had no series on the networks for a period following the cancellation of "The Wonderful World of Disney." Others on this list, however, such as GTG Entertainment, are new entrants. From 1984 to 1988, the number of producers identified by the *Almanac* as participants in prime-time series production ranged from a low of 30 in 1985 to a high of 45 in 1987. The turnover was substantial: 8 names were added and 3 dropped from 1984 to 1985, and the net gain of 4 producers from 1984 to 1988 is the product of 18 additions and 14 drops. Some of the disappearances undoubtedly represent mergers of preexisting network suppliers. Nevertheless, the turnover in the ranks of network suppliers is very high.

Even for those suppliers who are a constant presence on network television, year-to-year fluctuation in shares is considerable. Depending on which measure of shares we use, either 12 (Table 2.9) or 13 (Table 2.10) of the suppliers had maximum shares more than double their minimum shares during this five-year period. Only Warner Brothers did not have a maximum share at least 60 percent larger than its minimum share, and even for Warner Brothers the difference was more than a third.

Ease of entry and share instability are both consequences of the fundamental uncertainty associated with trying to predict viewers' preferences in programming from season to season. Uncertainty is greatest for new series, only about a quarter of which are renewed after their first season. The renewal rate is greater for series that have been renewed at least once, but approximately a quarter of older series are also dropped from network schedules from one year to the next.[25] Furthermore, for every series broadcast by the networks, a number of others were considered but did not make the final cut. The unpredictability associated with getting programs on the network schedules and keeping them there means that the fortunes of a given producer may rise or fall dramatically over a short period of time (see Tables 2.9 and 2.10). The unpredictability of demand for a program means that it is impossible for a producer or studio to plan with confidence on being able to keep a fixed stock of stages, equipment, talent, and other production inputs steadily employed with its own programs. In slow years the producer would have to absorb the costs of idle capacity and in hot years a prudent level of capacity would likely be inadequate. The risk-minimizing solution for dealing with the high uncertainty in program production has been to make all resources, including talent, available for hire. As fortunes change, capacity is reallocated to those who need it at the moment, and the availability of necessary resources on a for-hire basis makes it possible for established suppliers to expand or contract output rapidly. A rental market for the factors of production needed to create programs also limits the risk of entry for new suppliers, because they can quickly assemble the equipment, talent, and expertise required to begin production and because the sunk (nonrecoverable) capital costs of entering the business are relatively low.

Other Obstacles to Collusion

An industry may be structurally concentrated and hard to enter and yet be behaviorally competitive if firms in the industry find it difficult to coordinate their actions. Each member of a cartel always has a private incentive to cheat on an understanding with its rivals by lowering its own price slightly to take advantage of the greatly increased sales that would result. Collusion is easiest if products are homogeneous, with attributes that are easy to monitor. If the products of different firms are perfect substitutes, the only attribute that must be monitored is price. Even in this situation, cartel members may try to cheat with rebates and off-list sales that either are not disclosed or are hidden in side agreements for other products or services.

In program supply, no two products (programs) are the same. Casts must differ by necessity, and identical scripts are prohibited by copyright law. Also, there is no practical way to monitor the quality of inputs or the true terms at which program services are sold when studios and networks work so closely together to produce programs. There is an implicit increase in a network's payment to a producer, for example, if the network agrees to provide camera equipment that previously was supplied by the producer. Contracts are complex and include a variety of options for future services, some for future episodes of the same series, others for programs that might be spun off from the one under contract. Collusion is made all the more difficult by the practice of deficit financing, because it is impossible to figure out what portion of a deficit is a price concession to a network and not investment in future syndication earnings. Even if the industry that supplied programs to the broadcast networks were not structurally predisposed to be competitive, the nature of the product and contractual ambiguities would make effective collusion difficult.

The Supply of Syndicated Programs

Independent stations (stations that are not affiliated with networks) must get programs from non-network sources. Most of these programs are supplied by distributors, known as syndicators. They distribute programs on a station-by-station basis. Because the broadcast networks program only about two-thirds of the broadcast day, network affiliates must also get some of their programs from syndicators. Syndicators distribute off-network series and first-run syndicated programs. First-run syndicated programs are programs produced for non-network distribution. Most first-run syndicated programs are game and talk shows, but there has been an increase in situation comedies and dramatic series produced for syndication in recent years. Off-network series once dominated syndication. In the 1990s, however, first-run programs account for two-thirds or more of the audience for syndicated programs, a reversal we discuss in more depth in Chapter 5.

Suppliers of both types of programs are included in Table 2.11, which reports estimates of syndication revenues from sales to broadcast stations for the 1988–1989 television season by the largest distributors. Although the supply of syndicated programs is somewhat more concentrated than is the supply of programs to the networks, the HHI of 822 is still below the 1,000 threshold at which the antitrust authorities begin to take notice. If sales of smaller syndicators were included, the HHI would be even lower. The low level of concentration, combined with the product factors discussed above, suggests that the supply of syndicated programs is also competitive.

The Supply of Programs to Cable

Lack of data prevents us from providing concentration statistics on the supply of programs to cable, but we can say with some confidence that the supply also must be competitive. Many firms that supply the networks and syndicate programs to independent sta-

Table 2.11 The supply of syndicated programs, 1988–1989 television season

Syndicator[a]	Rank	1988–1989 Billings ($ millions)	Market share (%)
Warner	1	377	11.8
King World	2	370	11.5
MCA	3	347	10.8
Paramount	4	338	10.5
Columbia	5	304	9.5
Viacom	6	290	9.0
Fox	7	230	7.2
Turner	8	168	5.2
LBS	9	134	4.2
Buena Vista	10	115	3.6
Group W	11	110	3.4
Tribune	12	90	2.8
Orion	13	76	2.4
Multimedia	14	74	2.3
Worldvision	15	73	2.3
New World	16	66	2.1
MGM	17	45	1.4
Total		3,207	100
Hirschman-Herfindahl Index			822

Source: N. Weinstock, "Syndication's Hot 20 Outfits," *Channels,* February 1990, pp. 76–82.

a. Three firms identified as barter representatives for other companies were dropped from this listing because their sales were also counted as sales for the parent companies.

tions also supply programs to cable program services, and there are no important entry barriers for those that currently do not. In addition, there are a number of other suppliers, such as Turner Program Services, that have grown up with the cable industry. The product attributes that make difficult collusion among broadcast syndicators and suppliers to the broadcast networks should ensure that cable suppliers behave competitively as well.

Summary

The industry that supplies television programs is competitive. Basic structural factors are one reason. Entry is easy, the industry is unconcentrated, and given the unpredictability of viewers' preferences, the market shares of individual firms vary considerably from year to year. Program suppliers compete on the basis of their investments in program inputs (such as actors, directors, writers, editing, and special effects) that are difficult to monitor. In combination, these factors make it difficult, if not impossible, to raise program prices above a competitive level.

The competition among program suppliers and their relationships with customers are complicated by the durability of recorded programs. Some programs continue to draw an audience after they have been shown many times. Suppliers try to maximize the present value of the profits that accrue from the repeated showings of their programs by carefully controlling the sequence and timing of their releases into various distribution channels—broadcast networks, cable networks (pay, basic, and pay-per-view), independent stations, and video cassette and video disk outlets. These timed releases into different distribution channels are called windows. Windows differ in the profits they generate because distribution channels differ in their profit margin per viewer and in the size of the audiences they make available.

Each showing of a program reduces its value in subsequent releases, because the size of the audience willing to view it and the amount audience members are willing to pay for the privilege fall with each repetition of the program. Therefore, program suppliers try to schedule channels with higher per viewer profit margins before releases to channels with lower per viewer margins. The size of the audience reached by a channel also influences the timing of its window. If the per viewer profit margin is held constant, large audiences are more profitable than small audiences. The value of the profits generated by a channel falls the longer the delay until they are received, which creates an incentive to schedule channels that generate large audiences early in the release sequence. Because the value of profits received in the future declines more rapidly the higher the rate of interest, profit maximization dictates that schedules for releases into different windows be compressed when interest rates rise.

Windowing greatly complicates the competition among program suppliers. A program competes with other programs for access to the distribution channels in the windows into which it is released.

Because distribution channels compete with each other for viewers, it also competes indirectly with programs available concurrently through other distribution channels. Because a program may be viewed in one of its later release windows years after its initial release, old programs compete with newer programs produced by the same studios. Windowing affects production budgets as well. In deciding on a budget, a producer must consider the additional audience and revenue that may be generated in each of the windows in which a program will be released if the production budget is increased. The cumulative revenue in a number of windows will justify a larger budget than would be feasible if a program were released in just one or a few windows.

Sales to foreign markets are an important source of revenue for U.S.-produced programs and films. From the international trade in these video products, the United States realizes a substantial surplus. American productions dominate international trade in films and programs, and this success is largely due to the size and wealth of the English-speaking portion of the international film and program markets. The language advantage makes it profitable for American studios to produce much more expensive (and popular) programs and films than are produced in other countries. Windowing practices in foreign markets are similar to those in the United States.

3 / Traditional Models of Program Choice

How does a video store, eager to make money, decide which cassettes to stock? How does a television station choose what programs to broadcast? How does a cable television operator decide which cable networks to carry and how to bundle them together at various prices? And what is the effect of competition on these decisions? Two economic models, one developed in the early 1950s and the other in the mid-1970s, address these questions, not in terms of profitability, but in terms of policy implications. The models, and many elaborations on them, attempt to predict the diversity of programs offered and the likelihood that viewers will see the programs they prefer. Although the modelers were preoccupied with the extent to which the television industry performed in the public interest, their analyses depended on assessments of broadcasters' profits from different program strategies and the efficiency with which the television industry served viewers' interests in the programming it provided.[1] These traditional models should therefore be helpful to those who must make business decisions as well as to policymakers. More modern models are discussed in Chapter 4.

Critics of the television industry have charged repeatedly that there are too many mass-appeal programs and too few programs for viewers with specialized tastes. Complaints about lack of diversity peaked during the 1950s and 1960s—the pre-cable era.[2] When cable grew rapidly in the 1970s and 1980s, many anticipated a flowering of diversity on an abundance of cable channels. Although the number of programs has soared, critics of the industry claim that cable's potential for diversity has not been fulfilled and that the programs are just more of the same fare provided by broadcasters.

Should the television industry be criticized for lack of diversity any more than the magazine, book, and motion-picture industries are? If broadcasters do appeal primarily to mass tastes, what are the implications of this fact? If the television industry were structured differently, would programming for minority tastes be more likely to appear? Theories of program choice investigate these questions. Under various assumptions regarding viewer preferences, program costs, television technology, and industry structure, models predict program patterns and the efficiency with which viewers' demands for programs are satisfied.

To understand these models, three truths should be kept in mind. First, there is a sharp distinction between the behavior of a monopolist and the behavior of a competitive industry. Second, the spectrum of available programs is greatly affected by whether advertisers or consumers pay for the programs. Third, the structure of competition is much different when multichannel distributors, such as cable systems or retail video stores, are competing, than it is when single-channel distributors, such as television stations or broadcast networks, are competing.

Steiner Models

In 1952 Peter O. Steiner introduced a model of program choice that has been elaborated on since then by numerous economists. Steiner-type analyses have been enormously successful in academic and policy circles because of the richness of institutional detail that they can incorporate and the consistency of their results with the perceived failure of the advertiser-supported broadcast industry to satisfy consumers' diverse tastes.

Steiner's analysis and the many subsequent elaborations on it are extensions of the classic article on product competition by Hotelling (1929).[3] Hotelling argued that under specific demand assumptions two competing firms will market products of "excessive sameness." The result is analogous to what occurs under a two-party political system. If one candidate moves very far from center, the other candidate will move between him or her and the mass of voters. As a result, both candidates take a middle-of-the-road position. Hotelling's minimum differentiation result depends on assumptions more restrictive than he employed (D'Aspremont et al., 1979). However, his basic insights are still compelling and his spatial framework can be used to analyze many issues related to the economics of product variety.[4]

Steiner models posit institutional assumptions that are far more specific than those generally employed in the spatial literature. Nevertheless, there are numerous parallels in the two literatures, and two in particular are worth noting. First, Hotelling compared choices of production locations for rivalrous sellers to the choice of a monopolist. More general spatial analyses have shown that monopolists tend to differ from rivalrous sellers in the variety and number of products they offer, a generalization that holds in the Steiner literature. Second, in some circumstances there is no stable equilibrium with more than two sellers in a spatial market.[5] This possibility arises within the Steiner framework.

Steiner constructs viewer preferences by dividing the audience into subgroups. For example, an audience of 8,750 individuals could be classified into three groups, each with internally homogeneous tastes. Steiner then divides the programs into "program types." Assume that there are three distinct program types. Each viewer group ranks the three program types and a fourth option, nonviewing. Steiner's preferences for any given program type are exclusive—that is, in each case nonviewing is the second choice of an individual. A viewer is said to be "satisfied" if offered a program that he or she will view, and the measure for comparing alternatives becomes the number of satisfied viewers.

Viewer group	1	2	3
Group size	5,000	2,500	1,250
Viewer program preferences			
First choice	1	2	3
Second choice	nonviewing	nonviewing	nonviewing

Given these preferences, what programs will audience-maximizing broadcasters offer? (Broadcasters attempt to maximize audiences in Steiner's model because advertising revenue, which is the only type of revenue, is assumed to depend directly on audience size. The costs of programs are ignored.)

Now consider two ownership structures: monopoly and competition. Three television channels are available. In the monopoly all three channels are controlled by one owner who operates the channels under unified management. In the other structure each of the three channels is operated by a separate owner engaged in noncollusive rivalrous competition. Assume that where two programs of the same type are offered simultaneously, they share equally the total audience for that program type. What programs result under the two structures?

Program pattern under monopoly:
1 channel of program 1—5,000 viewers
1 channel of program 2—2,500 viewers
1 channel of program 3—1,250 viewers
Viewer satisfaction: 8,750 receive first choice

Program pattern under competition:
2 channels of program 1—2,500 viewers each
1 channel of program 2—2,500 viewers
Viewer satisfaction: 7,500 receive first choice
1,250 do not view

How did these program patterns come about? The monopolist, interested in increasing the total audience of the three channels, offers one program of each type. Each competing broadcaster is interested in maximizing the audience of its own channel. The first broadcaster to enter the market will show program 1, getting an audience of 5,000 viewers. The second broadcaster then has three choices: (a) show program 1, getting an audience of 2,500 (sharing group 1 equally); (b) show program 2, getting an audience of 2,500; or (c) show program 3, getting an audience of 1,250. Obviously, the second broadcaster will choose (a) or (b). Suppose the broadcaster chooses (b) and shows program 2. The third broadcaster can show program 1 (splitting group 1 with the first broadcaster), or program 2 (splitting group 2 with the second broadcaster), or program 3 (getting all of group 3). The third broadcaster will get the greatest audience if it duplicates program 1. Thus, under competition two broadcasters show program 1 and one broadcaster shows program 2. The arrangement is "stable," because none of the broadcasters can hope to gain by changing its program.

In this case monopoly yields greater viewer satisfaction than does competition. Groups 1 and 2 are indifferent to the program patterns, and group 3 clearly prefers that of monopoly. Competitors fail to offer the "minority" program, program 3. Monopoly yields greater "program diversity"; competitors engage in "program duplication," since two broadcasters simultaneously offer program 1. Steiner's model predicts that, under competition, product imitation will occur in the form of program duplication. This tendency toward excessive sameness is similar to that predicted by Hotelling, and it reaffirms what critics of the U.S. broadcasting structure claim to observe.

The result of program duplication in this case is not only forgone minority-taste programming, but a duplication of costs. Program duplication results in a waste of resources, as well as a potential displacement of minority programming. (In the real world, copyrights ensure that a broadcaster must offer a distinctly different program of the same program type. Hence a broadcaster must duplicate the production or purchase cost in order to engage in program duplication. Of course, pure duplication cannot occur; programs are close substitutes.)

Steiner (1952, p. 206) concludes that "a discriminating monopoly controlling all stations would produce a socially more beneficial program pattern" than would a competitive television industry. But note the strong assumptions that are required to produce this result: (1) viewer groups are highly unequal in size; (2) viewers watch only their first choices; (3) channel capacity is limited; (4) competitors duplicating a program share audiences equally; (5) all viewers are of equal value to broadcasters; (6) program costs are ignored. These assumptions will be relaxed in the model to follow. Steiner's results also depend on his measure of satisfaction. This will be true of any measure, but Steiner's is quite narrow.

In a study that complements Steiner's work, Rothenberg (1962) provides only two sketchy examples, but he focuses on two important variables: viewers' willingness to view programs other than their first choices, and the channel constraint. Rothenberg shows that if viewer groups each have a unique first choice, but as an alternative to nonviewing will watch some "common denominator," then competitors are likely to duplicate this program before providing preferred programs.[6] One can demonstrate this result by altering preferences slightly from those in the earlier example. The first program type is the common denominator.

Viewer group	1	2	3
Viewer program preferences			
First choice	1	2	3
Second choice	nonviewing	1	1
Third choice	nonviewing	nonviewing	nonviewing

Now note the result in the competitive market:

> Program pattern: 3 channels of program 1—2,917 viewers each
> Viewer satisfaction: 5,000 viewers receive first choice
> 3,750 viewers receive second choice

Under these conditions program duplication becomes even more attractive to competitors than Steiner had anticipated. Rothenberg concludes correctly, however, that competitive duplication is not a serious threat to viewer satisfaction if channel capacity is unlimited, because duplication in this case does not displace minority programs (although it may result in a waste of resources). Rothenberg does not consider the effects of monopoly ownership.

Wiles (1963) uses nine examples to predict program patterns under alternative ownership structures, channel capacities, and means of support. He shows program patterns to be highly sensitive to these factors. A problem with his analysis is that his predictions are sensitive to specific assumptions. Wiles (1963, p. 186) concludes: "Under the relationships between cost and revenue obtaining for television, minimum differentiation is most profitable to an oligopoly or monopoly, maximum differentiation to a polypoly" (a large number of channels in rivalrous competition). This statement needs to be examined carefully. Do oligopoly and monopoly necessarily produce minimum differentiation? In Wiles's examples monopoly produces maximum differentiation. Later, when Wiles states that television should be monopolized because the spectrum is limited, he appears to agree with Steiner.

The model of program patterns by McGowan (1967) differs from the other models. Each producer independently determines an optimal program mix for the entire "decision period." By failing to take into account the effect of one channel's program decision on another channel's audience, this model ignores the crux of small-group competition. McGowan also assumes away program duplication. As a result, his model generates unrealistic predictions: "Industry performance given the number of broadcast stations will, on the average, be the same when the several broadcast facilities are independently operated as it would be if they were operated under unified management" (McGowan, 1967, p. 512). By taking channel interdependencies into account, the Beebe model shows this conclusion to be contradicted in nearly all cases.

The Beebe Model

Steiner's model, in which the second choice of all viewers is nonviewing, allows an appealing satisfaction criterion: maximization of viewers for a given number of channels. However, for generalized preferences in which viewers rank several choices above nonviewing, this criterion is too narrow. It measures only the extent to which viewers are persuaded to view. A satisfaction measure that registers not only total audience but also the extent to which viewers receive their preferred choices would be more useful.

The culmination of Steiner's line of analysis is the simulation model developed by Jack H. Beebe. His more general model (1977) employs the following measures of viewer satisfaction. For each program pattern, the number of viewers receiving their first choice, second choice, and so on, or no program is tallied. Then the results under monopoly are compared with those under competition in terms of (1) which structure satisfies more first choices, (2) which structure attracts the larger audience, and (3) which structure wins a vote in which each viewer chooses the structure that gives him or her the higher-ranked program.

For the voting scheme, viewers who receive the same choice under the two structures are assumed to cast a vote of indifference. The voting scheme is a populist notion and not one necessarily favored by economists. Voting, however, provides a useful comparison of monopoly and competition, given that each member of the audience receives equal weight in the vote regardless of how intense his or her feelings are. Note that the results of the vote give no information as to which choices a program pattern is satisfying. The first two measures convey this information.[7]

In our example of Steiner's model, the program pattern under monopoly compares favorably with the pattern under competition. Under monopoly 8,750 receive their first-choice program compared with only 7,500 under competition. Monopoly attains a total audience of 8,750, while competition attracts only 7,500. Monopoly wins the populist vote because groups 1 and 2 are indifferent, while group 3 prefers monopoly.

A comparison of outcomes in terms of any of these measures does not avoid the problem of making interpersonal welfare comparisons. For example, competition may give 50 viewers their first choices, while monopoly gives first choices to only 30. But is competition better than monopoly in terms of first choices? There is no guarantee that all 30 viewers who receive their first choices under monopoly also receive first choices under competition. The 50 to receive first choices under competition might be an entirely different subset of the population. The measures count heads only. In this example, competition gives more first choices than does monopoly, but some viewer might still get a higher choice with monopoly.

To measure viewer satisfaction, one could assign dollar values to choices. For a given program pattern, viewer satisfaction would then be expressed in terms of a single index number, which would represent the value of the program pattern to viewers. Because viewers receive programs free under advertiser-supported television, the value of this index number would then become a measure of consumer surplus.[8] Although this measure of viewer welfare could easily be added to the model, it is not computed for two reasons.

First, all that is needed to predict program patterns under advertiser support is a much weaker assumption regarding the order in which viewers rank programs (an ordinal preference ranking). In the absence of reliable empirical measures of the dollar satisfaction that viewers derive from programs, the assignment of dollar values must be arbitrary. (It is preferable at this point not to incorporate subjective weights in the measure of broadcaster performance. These weights will be added later in the analysis of pay television. The pay-television analysis requires that viewer preferences be stated in dollar terms.)

The second reason for not expressing dollar satisfaction as a single index number is that the number would conceal information regarding which choices are being satisfied. For example, if a first choice is valued at 20¢, and a second choice at 10¢, does an index number representing 40¢ of total satisfaction mean that two persons received their first choices, or that four persons received their second choices? This detailed information will be valuable in analyzing the model's predictions.

To the extent that competitors duplicate programs and accordingly fail to offer unique programs, the effect on viewer satisfaction is recorded in the measures of choices. Since program duplication also implies duplicated resource costs, the total cost of duplication is not registered in the satisfaction measures. For this reason, the conditions under which program duplication occurs are discussed.

Assumptions

Beebe made the following assumptions in his analysis of advertiser-supported television:

1. The television market has a fixed number of potential viewers.
2. During a programming period of standard length, each channel presents a single program type. The model deals with only a single program period.
3. There are a finite number of identifiable and distinct program types, defined in terms of viewer preferences.[9]
4. The program is a free good to the viewer. Any cost of advertising paid through product prices is ignored, as is the opportunity cost of viewer time.
5. Viewers in the market can be classified into a finite number of internally homogeneous viewer groups.
6. Preferences are independent of the actual program types presented.
7. All channels simultaneously producing the same program type share equally in the total audience for that type.
8. The cost of a program is not affected by the actual number of viewers. In other words, the marginal cost of an additional viewer is zero.
9. Each program type has a given purchase cost associated with it. For a given program type, popularity is not a direct function of program cost. The relationship between program cost and popularity is handled by defining different program types and by varying the popularity across program types.
10. Advertising rates per viewer are fixed for the relevant decision period, and all viewers are worth the same amount to advertisers. Advertising revenues for a program depend upon the prevailing rate and the program's actual audience. (The assumption that all viewers are worth the same amount to advertisers is later relaxed.)

11. (a) For the monopolistic system, the operator seeks to maximize the joint profits of all channels. If any channel's operation results in a net loss to the operator, it has the option of discontinuing use of that channel with the resulting zero net profit from that channel. (b) For the competitive system, the operator of a single channel seeks to maximize its own profits in noncollusive rivalrous competition with other channels. The operator has the option of going off the air with a resulting zero net profit.

Calculations with the model are simplified by the fact that advertising rates per viewer are constant (assumption 10). Therefore, the fixed program costs (assumption 9) can be stated in terms of minimum break-even audiences. The profit-maximizing broadcaster will then attempt to maximize the difference between total audience and break-even audience. This difference is the measure of profit. Also assume that break-even audiences are equal for all program types (that is, program costs are the same for all program types). (Because popularities of programs will vary, this assumption is not too restrictive. It will be discussed later.) Under these conditions, the monopolist will maximize profits by maximizing the total audience of all television channels, provided that the net addition to audience from each channel is at least the break-even quantity. For the competitive system, a television channel operator will maximize its own audience, provided that this audience is at least the break-even quantity.

The break-even audience is held constant across program types, so that formally there is no correlation between program cost and popularity. However, there is symmetry between holding program popularity constant and varying costs and holding program costs constant and varying popularity. The model does the latter by varying the distribution of viewers. The effects of the correlation of program costs with popularity can be discussed, therefore, without introducing this relationship explicitly.

Within the framework of these assumptions, the model of advertiser-supported television uses a range of viewer preference structures and program supply characteristics. Within the range of these dimensions, the performance of a competitive broadcasting structure will be compared with that of a monopolistic structure.

Viewer Preferences

The model uses three patterns depicting viewer choices (see Table 3.1). In the first pattern each viewer group has a unique first choice and will watch only that program (Steiner's assumption). If offered any other program, the group will not view. The second pattern is more general. Each group still has the same unique first choice, but it will view an alternative program (closely related on the program spectrum) if its first choice is not available. If neither of these two programs is available, the group will not view. Additional lesser choices (third, fourth, etc.) can be added. However, a third case captures the effects of this expansion. Each group has a unique first choice and a closely related second choice, but in addition it will always view some common denominator rather than turn its sets off. Under this assumption, there exists a program that every group prefers to nonviewing. This case is suggested by Rothenberg and also is approximated by Hotelling's preference assumption.

These three patterns cover a wide range of possible preferences. The preference structure is not complete, however, until the number of viewers in each group is specified. Again a range of values is assumed. The hypothetical audience of 10,000 viewers is divided into a maximum of 25 groups according to three geometric distributions: a highly skewed distribution in which each successive group is only one-fifth as large as the previous group; a skewed distribution in which each group is one-half as large as the previous group; and a nearly rectangular distribution in which each group is nine-tenths as large as the previous group (see Table 3.1). A rectangular distribution in which all groups are of equal size could also be used. Many of the program patterns that result are indeterminate without arbitrary rules for broadcaster decisions when a number of programs offer equal profitability. The nearly rectangular distribution captures the flavor of the rectangular distribution but avoids indeterminacies. For the highly skewed distribution, mass audiences are very large relative to minority audiences. For the nearly rectangular distribution, all groups are of nearly comparable size.

Combining the three degrees of skewness in the viewer distribution with the three preference patterns yields viewer preferences that are very general. Since no one knows how preferences appear in reality, it becomes imperative to find predictions that hold under general preference assumptions.

Table 3.1 Beebe model's assumptions of viewer preferences and program supply

Preference Patterns

This is an ordinal ranking of program types by viewer groups. Choices of only the first five groups are shown.

1. Viewers will watch only their first choices.

#1 Programs	Viewers 1	2	3	4	5
1	1				
2		1			
3			1		
4				1	
5					1

2. Viewers have unique second choices.

#2 Programs	Viewers 1	2	3	4	5
1	1	2			
2		1	2		
3			1	2	
4				1	2
5					1

3. Viewers have a common lesser choice (the common denominator).

#3 Programs	Viewers 1	2	3	4	5
1	1	2	3	3	3
2		1	2		
3			1	2	
4				1	2
5					1

Viewer Distribution

Only the first five groups are shown.

A. Highly skewed distribution	B. Skewed distribution	C. Nearly rectangular distribution
Group Size	Group Size	Group Size
1. 8,000	1. 5,000	1. 1,077
2. 1,600	2. 2,500	2. 970
3. 320	3. 1,250	3. 872
4. 64	4. 625	4. 785
5. 12	5. 313	5. 707

Program Costs

a. Higher program costs (the break-even audience is 1,200 viewers).

b. Lower program costs (the break-even audience is 800 viewers).

The Channel Limitation

a. Channel capacity limited (three channels).

b. Channel capacity unlimited.

Program Supply

The model includes two critical dimensions of program supply: the cost of purchasing programs and the number of television channels available in the market. The explicit inclusion of program costs is important for two reasons. First, in a model that postulates profit maximization, programs must be economically viable. It is irrelevant to argue for minority programming in a purely profit-maximizing context without specifying the cost of such programs. Second, by altering assumptions regarding program costs, one can judge the effects of program costs on program patterns and viewer satisfaction.

In Table 3.1 two levels of break-even audience size are used as proxies for higher and lower program costs. The number of channels is assumed to be either limited (arbitrarily to three or six channels) or unlimited. Given the earlier assumptions about channels and costs, it is possible to calculate the program patterns that result from any particular combination of assumptions.

Program Patterns and Viewer Satisfaction

Five important cases generated by the Beebe model are presented in this section. The first four cases examine Steiner's conclusions. Case 1 shows that, under Steiner's assumptions, monopoly provides greater viewer satisfaction than does competition—Steiner's result. Cases 2 and 3 show that if additional channels are allowed, or if viewer groups are nearly equal in size, monopoly and competition provide the same viewer satisfaction. Case 4 shows that if viewers will watch other than their first choices, they may prefer competition to monopoly—the opposite of Steiner's result. Case 5 then demonstrates that program patterns may be unstable under competition. These five cases are representative of the wider range of cases presented later in Tables 3.2 and 3.3.

Steiner's conclusion (1952, p. 206) that under limited channels "a monopoly controlling all stations would produce a socially more beneficial program pattern" can be examined using the simulation model and the dimensions of preferences and program supply in Table 3.1. First, consider case 1, which is analogous to Steiner's.

Under competition, broadcasters in case 1 find it more profitable to engage in program duplication than to offer a diversity of programs. The monopolist, however, offers all programs that are economically viable. Monopoly clearly produces more viewer satisfaction (without incurring the cost of program duplication) than does competition. Can one conclude, as did Steiner, that viewers would prefer all stations to be operated under unified management? This conclusion is sensitive to three assumptions: the number of channels is limited; the viewer distribution is highly skewed; and viewers will watch only their first choices. Case 2 relaxes the assumption that channels are limited.

When the number of channels is expanded to six, as it is in case 2, the monopolist leaves the additional channels dark. Competitors fill the additional channels by continuing to duplicate program 1, further splitting the mass of viewers. Given enough channels, it finally becomes profitable for competitors to offer program 2. Given ample channels, minority programs (if economically viable) eventually appear under competition in spite of program duplication. Under unlimited channels, competitive duplication of programs causes no loss in viewer satisfaction (although it wastes resources). Even under Steiner's strict preference assumption, if there are unlimited channels, monopoly no longer provides superior choices.

In case 3, where the preference assumption is relaxed, competitors no longer engage in program duplication. Program duplication

Case 1 Steiner's case

---

a. Preference assumption 1A (viewers will watch only their first choices and the viewer distribution is highly skewed).
b. Break-even audience is 800.
c. Number of channels limited to three.

Monopoly

Program pattern: 1 channel of program 1—8,000 viewers
1 channel of program 2—1,600 viewers
1 channel dark
Viewer satisfaction: 9,600 receive first choice
400 do not view

Competition

Program pattern: 3 channels of program 1—2,667 viewers each
Viewer satisfaction: 8,000 receive first choice
2,000 do not view

Case 2 Additional channels

a. Preference assumption 1A (same as before).
b. Break-even audience is 800 (same as before).
c. Six channels in the market.

Monopoly

Program pattern: 1 channel of program 1—8,000 viewers
1 channel of program 2—1,600 viewers
4 channels dark
Viewer satisfaction: 9,600 receive first choice
400 do not view

Competition

Program pattern: 5 channels of program 1—1,600 viewers each
1 channel of program 2—1,600 viewers
Viewer satisfaction: 9,600 receive first choice
400 do not view

Case 3 Viewer distribution no longer skewed

a. Preference assumption 1C (viewers will watch only their first choices and the viewer distribution is nearly rectangular).
b. Break-even audience is 800 (same as before).
c. Number of channels limited to three.

Monopoly

Program pattern: 1 channel of program 1—1,077 viewers
1 channel of program 2—970 viewers
1 channel of program 3—872 viewers
Viewer satisfaction: 2,919 receive first choice
7,081 do not view

Competition

Program pattern: 1 channel of program 1—1,077 viewers
1 channel of program 2—970 viewers
1 channel of program 3—872 viewers
Viewer satisfaction: 2,919 receive first choice
7,081 do not view

under competition results when a mass of viewers who prefer one program is considerably larger than another mass who prefer a different program. But this occurs only if viewer groups are skewed in size (or if a common denominator exists). If duplication does not occur and if viewers will watch only their first choices, then the two ownership structures provide identical program patterns. Again, monopoly does not necessarily provide superior satisfaction to viewers.

As long as viewers refuse to watch any programs except their first choices, then the model predicts that monopoly always provides viewer satisfaction at least as great as that provided under competition. However, viewers may watch less-preferred choices if their first choices are not available. Consider cases 4 and 5.

Where viewers will watch lesser choices, the monopolist will offer only audience-maximizing common denominators. Why should the monopolist produce program 2 (the second group's first choice) when it can capture group 2 by offering only program 1? The monopolist's only concern is that viewers choose television over other activities, not that they watch their favorite programs.

Case 4 Viewers will watch lesser choices

---

a. Preference assumption 2B (viewers will watch only their first and second choices and the viewer distribution is skewed).
b. Break-even audience is 800 (same as before).
c. Number of channels limited to three (same as before).

Monopoly

Program pattern: 1 channel of program 1—7,500 viewers
1 channel of program 3—1,875 viewers
1 channel dark
Viewer satisfaction: 6,250 receive first choice
3,125 receive second choice
625 do not view

Competition

Program pattern: 2 channels of program 1—2,500 viewers each
1 channel of program 2—3,750 viewers
Viewer satisfaction: 7,500 receive first choice
1,250 receive second choice
1,250 do not view

Yet to gain audiences, competitive broadcasters must attract viewers away from other channels as well as away from other leisure activities. Therefore, competitors find it profitable to offer preferred choices. In case 4 the competitor offers program 2, thereby attracting group 2 away from the other channels that offer program 1. (If there were more channels, the competitor also would offer program 3. The monopolist, however, would never offer program 2.) The comparison of monopoly and competition looks very different when lesser choices are allowed. No longer do the monopolist's program patterns provide viewer satisfaction at least as great as the competitor's program patterns. (Under preference pattern 3, the monopolist always offers only the common denominator, regardless of other variables.)

In case 4 competition satisfies more first choices, while monopoly attains a larger audience. Therefore neither structure is strictly preferred by viewers. (Competition wins the populist vote, however.) Group 2 votes for competition, groups 3 and 4 vote for monopoly, and all other groups are indifferent.)

Case 5 demonstrates that where lesser choices exist, an equilibrium in pure strategies may not always occur under competition. Instead, a mixed strategy or probabilistic outcome occurs.[10] There is no reason to suppose that program patterns should be fixed or stable. Chamberlin (1956) suggests this result in his discussion of Hotelling's model. Indeed, rivalrous networks change their program strategies continuously.

The game used in the simulation model to produce the competitive program patterns is an *n*-person, non-zero-sum noncooperative game, where *n* is sufficiently small that one channel's program decision directly affects profits on other channels. Each producer is treated symmetrically; that is, each competitor is allowed the same payoff from a given strategy. A solution in pure or mixed strategies is sought using a method of fictitious play.

Consider how the mixed strategy comes about. First suppose the three competitive broadcasters offer programs 1, 2, and 3, respectively. In this situation it obviously pays the producers of program 1 or 2 to shift to program 4, thereby attaining a net gain in audience. If this occurs, then it pays the producer of program 3 to shift to program 1 or 2. But given this move, it pays the producer of program 4 to shift to program 3, and so forth. The three producers will continue to shift among the four programs. No stable pattern of

Case 5 Competitive mixed-strategy equilibrium

a. Preference assumption 2C (viewers will watch only their first and second choice and the viewer distribution is nearly rectangular).
b. Break-even audience is 800 (same as before).
c. Number of channels limited to three (same as before).

Monopoly

Program pattern: 1 channel of program 1—2,047 viewers
1 channel of program 2—1,657 viewers
1 channel of program 5—1,343 viewers
Viewer satisfaction: 2,656 receive first choice
2,391 receive second choice
4,953 do not view

Competition

Program pattern: Three channels in a mixed strategy such that programs 1–4 are each shown with 75 percent probability. The expected (and average) audience for each channel is 1,322 viewers.
Viewer satisfaction (mixed strategy averages):
2,778 receive first choice
1,187 receive second choice
6,035 do not view

pure strategies exists, and the result is a mixed strategy for each producer. (For the mixed strategy, the resulting program patterns and viewer satisfaction are calculated in probabilistic terms for the cycle over which the mixed stategy occurs.)

In case 5 as in case 4, it is not clear which structure provides preferred choices. Competition satisfies more first choices, but monopoly gains a greater total audience. (In case 5, monopoly wins the populist vote.)

The 18 cases in Tables 3.2 and 3.3 are generated across the spectrum of the three preference patterns, three viewer distributions, two break-even audiences, and two channel options given in Table 3.1. Table 3.2 provides predictions of the highest-numbered program type offered by monopoly and competition. This is a measure of programming for minority-taste audiences, because higher-number programs, by convention, are always those aimed at smaller audiences. The table also gives the number of channels used, since this is a measure of the spectrum crowding that is neces-

Table 3.2 Highest numbered program type offered and number of channels used under monopoly (Mon.) and competition (Comp.)

Case	Preference pattern[a]	Viewer distribution[a]	Minimum audience	Three channels: Highest numbered program type offered		Three channels: Number of channels used		Unlimited channels: Highest numbered program type offered		Unlimited channels: Number of channels used	
				Mon.	Comp.	Mon.	Comp.	Mon.	Comp.	Mon.	Comp.
1	1	A	1,200	2	1	2	3	2	2	2	7
2			800	2	1	2	3	2	2	2	12
3		B	1,200	3	2	3	3	3	3	3	7
4			800	3	2	3	3	3	3	3	10
5		C	1,200	0	0	0	0	0	0	0	0
6			800	3	3	3	3	3	3	3	3
7	2	A	1,200	1	1	1	3	1	2	1	7
8			800	1	1	1	3	1	2	1	12
9		B	1,200	3	2	2	3	3	3	2	7
10			800	3	2	2	3	3	4	2	11
11		C	1,200	5	4[b]	3	3	5	6[c]	3	5–6[d]
12			800	5	4[b]	3	3	9	9[c]	5	9–10
13	3	A	1,200	1	1	1	3	1	2	1	7
14			800	1	1	1	3	1	2	1	12
15		B	1,200	1	2	1	3	1	3	1	7
16			800	1	2	1	3	1	4	1	11
17		C	1,200	1	1	1	3	1	6[c]	1	9–10[e]
18			800	1	1	1	3	1	9[f]	1	13–15[e]

a. See Table 3.1
b. The probability that this program appears in the mixed strategy is 75%.
c. The probability that this program appears in the mixed strategy is 96%.
d. The range is taken over the mixed strategy.
e. For these cases the occurrence of a mixed strategy under the Cournot assumption leads to losses for some broadcasters.
f. The probability that this program appears in the mixed strategy is 92%.

Table 3.3 Viewer satisfaction under monopoly and competition (M = monopoly satisfies more viewers; C = competition satisfies more viewers; E = the two structures satisfy equal numbers of viewers)

				Three channels			Unlimited channels		
Case	Preference pattern[a]	Viewer distribution[a]	Minimum audience	First choices	Total viewers	Populist vote	First choices	Total viewers	Populist vote
1	1	A	1,200	M	M[b]	M	E	E[b]	E
2			800	M	M	M	E	E	E
3		B	1,200	M	M	M	E	E	E
4			800	M	M	M	E	E	E
5		C	1,200	E	E	E	E	E	E
6			800	E	E	E	E	E	E
7	2	A	1,200	E	E	E	C	C	C
8			800	E	E	E	C	C	C
9		B	1,200	C	M	C	C	E	C
10			800	C	M	C	C	C	C
11		C	1,200	C	M	M	C	C	C
12			800	C	M	M	C	M[c]	C
13	3	A	1,200	E	E[d]	E	C	E[d]	C
14			800	E	E	E	C	E	C
15		B	1,200	C	E	C	C	E	C
16			800	C	E	C	C	E	C
17		C	1,200	E	E	E	C	E	C
18			800	E	E	E	C	E	C

a. See Table 3.1.

b. For preference pattern 1, viewers receiving first choices are identical to total viewers.

c. Monopoly attains a slightly larger audience because the monopolist always shows program 9, but it is shown with 96 percent probability under competition (see Table 3.2). This is a peculiarity of the simulation model that can occur under a mixed strategy because audience groups are lumpy in size.

d. For preference pattern 3, as long as program 1 is shown the entire audience will view.

sary to obtain this extent of minority programming. Table 3.3 compares monopoly and competition in terms of first choices, total audience, and a populist vote between the two structures.

Monopoly and Competition

Two important behavioral differences between monopolists and competitors influence program patterns and viewer satisfaction: competitors tend to duplicate and imitate programs, and monopolists search for common denominator programs to minimize program costs.

Competitive duplication results under the skewed viewer distribution (A or B in Table 3.1) and under common denominators (preference pattern 3). Duplication exemplifies the Steiner criticism of competition. The monopolist's behavior is the result of protection from product competition. The monopolist never duplicates programs and under advertiser-supported television avoids producing substitutes. If viewers accept other than their first choices (preference patterns 2 or 3), then the monopolist specifically seeks out common denominators.

Where viewers prefer some common denominator to nonviewing (preference pattern 3), this is the only program the monopolist will offer (see Table 3.2). Competitors are likely to duplicate this program, but given ample channels (and sufficient demand for preferred types to cover costs) sufficiently numerous competitors eventually cater to higher choices. Under monopoly, if viewers expect preferred choices, they must exercise their power of nonviewing when offered lesser choices. But under competition with unlimited channels, viewers' likelihood of receiving preferred choices is independent of whether they rank lesser choices.

Program Costs

Steiner's model ignores program costs and hence assumes implicitly that channel capacity is the only binding constraint. The implication is profound: whenever program duplication exists, it displaces minority-taste programs. But when program costs are introduced, there exists the possibility of excess channel capacity. Under excess channel capacity, duplication does not displace minority programming but appears in addition to it and thus causes no loss in viewer satisfaction. Although duplication wastes resources, viewers are not denied alternative programs.

The displacement of minority-taste programs (which occurs only under limited channels) is probably a more serious cost of duplication than is the cost of duplicated inputs. Is duplicated editorial effort in news or fashion magazines a competitive waste of resources? In reality, are programs or magazines ever perfect substitutes? Might the threat of duplication improve quality? Duplicated editorial effort in magazines might pose a serious problem if only three magazines were available to provide diverse content.

Channel Capacity

If one wishes programs for a minority-taste audience to be available, then adequate channel capacity is a necessary but not a sufficient condition under profit-maximizing ownership and advertiser support (see Table 3.2). Sufficient conditions under competition are (1) adequate channel capacity, and (2) some audience subset greater than the break-even audience size that ranks the program as a first choice. Sufficient conditions under monopoly are the above plus (3) the group's refusal to view less-preferred programs (preference pattern 1). The addition of channels, *ceteris paribus,* never leads to a decrease in program offerings or viewer satisfaction.

Under unlimited channel capacity, the number of channels used varies considerably depending upon assumptions about preferences, program costs, and ownership structure. This suggests that there is no way to know a priori what will be an adequate number of channels for a cable television system, for instance, without specifying preferences, program costs, and institutional structure as well as the cost of additional channels.

Policy Choices with Limited Channels

Under limited channels, the Beebe model yields the following three predictions. First, if viewers watch only their first choices (preference pattern 1), then monopoly provides at least as many program types and viewer satisfaction at least as great as competition. Second, if viewers watch only their first choices, and the viewer distribution is skewed (viewer distributions A or B)—which results in program duplication under competition—then monopoly provides more program types and greater viewer satisfaction than does competition. Third, if there is a common denominator (preference pattern 3), then competition provides at least as many program types and viewer satisfaction at least as great as monopoly (which provides only the single program).

Steiner and Wiles each conclude that if channel capacity is limited, then monopoly will likely produce more program types and yield higher viewer satisfaction than will competition. This prediction is necessarily true only under the strong assumptions of the second prediction in the preceding paragraph. Under more general (and realistic) preferences, the conclusion of Steiner and Wiles is not necessarily true. Under general preferences and limited channels, one cannot say (without stating specific preference assumptions) which ownership structure offers more program types and yields higher viewer satisfaction (see Tables 3.2 and 3.3, columns under "Three Channels"). If viewers will watch only their first choices, then monopoly is likely to be the preferred structure. If viewers will watch lesser choices, and viewer groups are not highly skewed in size, then competition is likely to provide more first choices, but monopoly is still likely to attract a larger audience.

Policy Choices with Broadband Technology

Dramatic increases in the number of channels delivered by broadband (multichannel) media such as cable and satellite television systems have made the unlimited channels version of Beebe's model more relevant now than when Beebe developed it. Although no medium offers a literally unlimited number of channels, channel capacities have grown to the point that unlimited channel capacity is a reasonable simplifying assumption for applied program choice analyses.

Under unlimited channels, competitive program duplication does not displace minority programs. The minority programs (if preferred by viewers and if economically viable) appear in addition to the duplicated programs (see Table 3.2, columns under "Unlimited Channels"). Under unlimited channels, therefore, competitive program duplication causes no loss in viewer satisfaction. In their attempts to attract audiences from other channels, competitors offer all economically viable programs, so viewers are guaranteed that they will see preferred choices (as long as the number of viewers preferring a given program type is large enough to cover the cost of making the program).

Under unlimited channels, the monopolist still seeks out common denominators (see Table 3.2, columns under "Unlimited Channels"). If the monopolist can capture an audience with a lesser choice, it has no incentive to provide a preferred choice, even though channel capacity is available.

Because of these results for competition and monopoly, the model predicts the following: Under unlimited channels, competitive ownership of channels always results in at least as many program types—and where lesser choices are specified, more program types—as does monopolistic ownership. This does not necessarily mean that competition always offers the same program types as monopoly plus additional types. Rather, it implies that the total number of different program types being offered under competition will exceed that under monopoly. Competition does not always offer the same program types as monopoly plus additional types because in the mixed strategy, it is not certain that programs offered by the monopolist will be offered by competitors. Case 5 provides an example.

Competition never satisfies fewer first choices than does monopoly (see Table 3.3, columns under "Unlimited Channels"). Where viewers will watch lesser choices, competition always satisfies more first choices. Where viewers will watch only their first choices, the two structures satisfy exactly the same number of viewers.

Competitors under unlimited channels will always satisfy economically viable preferred choices (although they may not offer these programs with certainty). This is not true of the monopolist under unlimited channels. Furthermore, under unlimited channels, competitors will almost always attain a total audience as great as that of the monopolist (see the last column in Table 3.3).

The Beebe model's predictions point to these conclusions. Under advertiser-supported television, if channel capacity is constrained (the present situation for VHF), then one cannot say which ownership structure will yield greater viewer satisfaction for general preferences. If channel capacity is unconstrained (which may effectively be the case with broadband video services), then the competitive structure will yield greater viewer satisfaction for general preferences (although costs attributable to program duplication will occur under the competitive structure).

Relaxing Two Assumptions

The model's assumptions 7 and 10, that channels that duplicate a program share equally in the audience for that program and that all viewers are equally valuable to advertisers, can now be relaxed. The first assumption will be relaxed because program duplication is an important result in the model, and the degree of program duplication under competition obviously depends on the way broadcasters duplicating a program actually divide (or expect to divide) the audience. It is important to examine the second assumption because advertisers may be willing to pay more (or less) for various minority-taste audiences. If this is the case, then the quantity of minority programming will be affected.

To relax the first assumption, consider what happens to program duplication if broadcasters do not split the audience equally. Suppose that each successive broadcaster duplicating a previously offered program receives half of the audience of the preceding broadcaster. For example, two broadcasters split the audience $2/3$ and $1/3$; three broadcasters split $4/7$, $2/7$, $1/7$. (In deciding to duplicate a program, the entering broadcaster assumes that it will obtain a fraction of the total audience for that program equal to $1/(2n - 1)$, where n is the number of broadcasters duplicating the program, including himself.) Consider again the case 1 preference assumptions, which were used to exemplify program duplication under competition.

Viewer group	1	2
Group size	8,000	1,600
Viewer program preferences		
First choice	1	2
Second choice	nonviewing	nonviewing

What program patterns emerge under the competitive broadcast structure with three channels, given the alternative assumptions regarding how broadcasters split audiences?

Assumption (a): Broadcasters duplicating a program share audiences equally.

Program Pattern (a): 3 channels of program 1—2,667 viewers each

Assumption (b): Each successive duplicating broadcaster receives half the audience of the preceding broadcaster.

Program Pattern (b): 2 channels of program 1—5,333 and 2,667 viewers respectively
1 channel of program 2—1,600 viewers

Where broadcasters duplicating a program do not share the audience equally, less duplication results. Duplication under the competitive structure is sensitive to the way in which broadcasters split audiences. At one extreme, a broadcaster who considers duplicating a program might predict that it can take the entire audience from the other broadcaster(s). This assumption leads to maximum duplication (and an obvious instability in program patterns). At the other extreme, a broadcaster who considers duplicating a program might think that the entire audience will remain loyal to the previous broadcaster. This assumption leads to no duplication. The model's assumption that broadcasters duplicating a program split the audience equally is admittedly arbitrary. As long as it is possible for an imitative broadcaster to obtain some fraction of the program's audience, however, then program duplication may occur under competition. In this case, the qualitative results of the model are independent of the actual shares assumed, even though the quantitative degree of program duplication is sensitive to this parameter.

Now let us relax the assumption that all viewers are equally valuable to advertisers. Suppose that incomes, tastes, and expenditure patterns differ across viewer groups. As a result, advertisers value some groups more highly than others. (Each individual advertiser will have its own valuation of viewer groups. But here we are concerned with the net aggregate effect of individual advertiser demands, because that is how the advertising market as a whole values each viewer group.) Assume that either (1) the smaller audience groups (the minority audiences) are worth more per viewer in the advertising market than are the larger (mass) audiences, or (2) the opposite is true. Which of these assumptions is more realistic is not obvious.

In generalizing the model of advertiser-supported television, assume first that all viewers will watch only their first choices (preference pattern 1). Assume also that advertising rates (per viewer) accurately reflect the values of viewer groups in the advertising market. Then the potential revenue to a broadcaster attracting any particular group is the number of viewers in the group weighted by the group's value per viewer to advertisers. This case 1 viewer preference structure is as follows.

Viewer group	1	2
Group size	8,000	1,600
Viewer program preferences		
First choice	1	2
Second choice	nonviewing	nonviewing

But now assume that different viewer groups generate different revenues per viewer. One can no longer use the concept of audience maximization; instead, one must revert to the more fundamental assumption of profit maximization (assumption 11). Assume exposure to viewers in group 1 is worth 1¢ per viewer and exposure to group 2 is worth 2.5¢ per viewer. (For completeness, the minimum audience sizes used earlier must also be converted to costs. For the minimum audiences of 1,200 and 800, substitute $12 and $8 as break-even costs. For the simple examples in this section, however, these costs are not needed.) Then the revenues generated from programs 1 and 2 are:

Program 1	8,000	×	$.01	=	$80.00
Program 2	1,600	×	$.025	=	$40.00

Based on this revenue structure across program types (and taking into account the break-even costs), profit-maximizing program patterns can be calculated for the model just as they were earlier. Note for this example that the audience for program 2 generates exactly half of the revenue that the audience for program 1 generates. (Program 3 would generate half the revenue of program 2, and so forth.) The relative revenues across program types are identical to those of the original model assuming the second viewer distribution (that is, a viewer distribution of 5,000, 2,500, 1,250, etc.). The predictions that result from the model are also identical. Therefore, one can interpret the effect of viewer groups' being worth differing amounts within the context of the original model merely by altering the viewer distribution. (This is strictly true where viewers watch only their first choices.) The result is that if minority audiences are worth more per viewer to advertisers, then this is equivalent to assuming a less skewed viewer distribution in the original model, with the same effect on program patterns. Minority programs are more likely to appear under both competition and monopoly, and program duplication is less likely to occur under competition. If mass audiences are worth more per viewer to advertisers, then the opposite will hold true.

For preference patterns in which viewers rank more than a single choice, calculating the revenues received by broadcasters and the resulting program patterns is more complicated. However, the predictions are basically the same as those that hold for the simpler case in which viewers watch only their first choices. When advertisers are willing to pay more for exposure to certain viewers, then the viewers who are worth more to advertisers will receive heavier weighting in the calculation of potential advertising revenues.

Does this mean that the groups that are worth more to advertisers are more likely to see the programs they want? The answer is yes. But does it also mean that viewers in general are more likely to see their preferred programs? No, or at least not necessarily. The signals that enter broadcasters' profit calculations and determine program patterns are the advertisers' values of exposure to viewers, not the viewers' values of programs. Prices in advertising markets reflect the amounts that advertisers are willing to pay for different viewers, but this does not necessarily mean that program patterns will more nearly reflect those desired by viewers.

If mass audiences are worth more per viewer than are minority audiences in the advertising market, then this will cause less diversity and more duplication of programs than predicted by the model of advertiser-supported television, with all viewers worth equal amounts. It seems reasonable, however, that as the number of channels continues to increase, advertisers who seek out minority groups will find the television medium increasingly attractive as an advertising vehicle whether or not these viewers are worth inherently more than mass audience viewers. In the early 1990s the trend is apparent in the proliferation of specialized cable services.

Because they serve specific needs of viewers and advertisers, these services are supported by a combination of viewer payments and advertising revenues. But note two important points: additional channels are a necessary condition for specialized programs to appear, and program patterns must still reflect advertisers' values of exposure to viewers in addition to viewers' values of the programs.

Pay Television

In the description of the Beebe version of the Steiner model up to this point, programs have been assumed to be supported wholly through advertising. Now consider the case of pure subscriber-supported television (that is, pay television in which the only revenues broadcasters receive are those that viewers pay on a per program basis). Broadcasters purchase programs in the programming market; they then sell these programs directly to viewers.

The measure of viewer satisfaction, like that used for the model under advertiser support, is the extent to which viewers receive their preferred choices (specifically, first choices). The program is a public good (a good whose cost of production is independent of the number of persons who consume it). One person's consumption of the good does not reduce the availability of the good for another person's consumption. Under advertiser-supported television, programs are used to attract audiences, and then the audiences are sold to advertisers. Under pay television, programs are sold directly to viewers, but programs still are public goods.

The program is a public good regardless of whether viewers receive it free or must pay for it. Being able to exclude viewers by enforcing payment does not convert the public good to a private good. Furthermore, the program is a public good whether the medium is over-the-air broadcasting (a public good), cablecasting (closer to a private good), or a video cassette (a pure private good).

Advertiser-supported television leads to an inefficient allocation of resources so long as advertisers' values of viewers differ from viewers' values of programs. The theory of public goods provides a basis for expecting competitive program duplication to be less serious under pay television than under advertiser-supported television. (On the efficiency of competitive markets in public goods, see Thompson, 1968; Demsetz, 1970; and Borcherding, 1978. Samuelson [1954, 1955, 1958] derives the conditions for the efficient allocation of a public good.)

Consider the dilemma that is posed by private (free-enterprise) production of a public good. Once a program is created, the marginal cost of having an additional person receive it is limited to the transmission expense of reaching that extra person. If the price charged exceeds this incremental transmission cost and thereby excludes the person, then society has lost some costless benefit. Unfortunately, at a zero price above the incremental transmission cost, there will be no incentive for the private production of programs because the fixed cost of producing the program will not be covered. This is a dilemma of some importance. The existence of a market for the public good depends on the presence of potentially inefficient exclusion in the market.

Under conditions of decreasing costs per consumer, one cannot expect the entrepreneur to price at the marginal cost of serving the consumer. The (fixed) cost of production would not be covered, and at this price the producer would have no incentive to produce the public good. As long as consumers do not all value the public good identically, there may not exist a uniform charge for the good's consumption that brings forth the efficient level of the public good.

One way out of this dilemma is to allow the firm to practice price discrimination among consumers, charging each the highest price he or she will pay for the public good. Theoretically, such a price-discriminating producer could produce an efficient allocation of the public good. Each consumer would pay a price equal to his or her true value of the good, and no consumer would be excluded. Such a scheme could satisfy Samuelson's conditions for the efficient provision of public goods. For this reason, windowing, as discussed in Chapter 2, may make television markets more efficient.

The Behavior of Competitors

In the model of advertiser-supported television, the price per viewer that accrued to broadcasters was determined in the advertising market. Competitive broadcasters attracted viewers through product competition, not price competition. Consequently, programs were duplicated. Under pay television, however, broadcasters will compete for viewers through product and price competition. Does the addition of price competition significantly alter the extent of competitive program duplication?

Suppose broadcaster A purchases program A at a cost of $400 and sells it to 10,000 viewers at 10¢ each. Its profit is $600. Broadcaster B now considers duplicating program A at a cost of $400, splitting the audience at 10¢ each, thereby earning a profit of $100 while reducing A's profit to $100.[11] As B contemplates duplicating program A, what reaction might B expect from broadcaster A? Because A has already incurred the $400 fixed cost, A will threaten (or carry out) price cuts in order to deter B from entering its market. If one ignores costs of transmission and billing, then broadcaster A will threaten price cuts to the point at which price equals marginal cost, which equals zero. The expected price competition from broadcaster A will render B much less enthusiastic about the prospect of duplicating program A than would be the case if there were no price competition.

This example assumes that broadcaster A has already incurred the $400 program purchase cost and is serving the market prior to B's appearance. But let us assume that both A and B are simultaneously contracting with viewers to determine which broadcaster viewers will buy from, and that neither broadcaster has yet incurred the $400 program cost. (This is similar to the mechanism considered by Demsetz, 1970.) In competing for the market, only one broadcaster will end up offering the program. (It is entirely possible that no broadcaster will produce the program due to the uncertainty of an additional entrant and the almost certain price war that would ensue if the second broadcaster did enter. The point, however, is that only one broadcaster, at most, will enter.) Furthermore, the price charged will be "competitive" (in this case close to 4¢ per viewer), because the broadcaster who "wins the contract" does so by offering the program at the lowest possible price. Even if firms do not contract with viewers in advance, a single broadcaster serving a market may still be forced to charge the competitive price if faced with the threat of entry by other broadcasters that might offer a lower price.

Here the price is determined in a competitive framework, whereas in the example above it was assumed to be 10¢.[12] The result regarding program duplication is the same, however. When price competition is a real possibility, program duplication (that is, the duplication of perfect-substitute public goods) does not occur. The qualification concerning perfect substitutes is an important one. As we shall see in Chapter 4, if programs are close but not perfect substitutes, even with pay support, too many mass-appeal programs and too few minority-taste programs will be produced. This tendency is much more pronounced with advertiser support.

The Behavior of a Monopolist

Suppose that all channels in a pay-television market are operated under unified management by a pay-television monopolist. (Perhaps the monopolist is the owner of the local cable television system.) A conclusion of the model under advertiser support was that where viewers were worth equal amounts in the advertising market, the monopolist sought common denominator programs. As long as it could capture a viewer, the monopolist had no incentive to provide the viewer with a preferred choice. Does this still hold for the monopolist in a pay-television market?

Viewers are presumably willing to pay higher prices to view their preferred choices. Under pay television, unlike advertiser-supported television, the potential revenue per viewer is higher for preferred choices. Therefore, the pay-television monopolist will no longer produce only common denominator programs. Consider a simple example with only two viewer groups and two program types. Viewers in group 2 will pay 10¢ to view program 1, but will pay 20¢ to view their preferred program, program 2.

Viewer group	1	2
Group size	5,000	2,500
Amounts viewers will pay for programs		
Program 1	10¢	10¢
Program 2		20¢

If programs cost $200 to purchase (or produce), then the monopolist offers both program 1 (at a price of 10¢) and program 2 (at a price of 20¢), thereby maximizing its profit at $600. (The marginal profitability of adding program 2 is 20¢ × 2,500 − 10¢ × $2500 − $200 = $50. The profit would be $550 if the monopolist offered only program 1.) Because viewers will pay more for preferred choices, the monopolist in this example caters to preferred choices.

This never occurs in the model under advertiser support, where viewers are worth identical amounts in advertising markets.

Now suppose that viewers in group 2 are willing to pay only 14¢ for their preferred choice. Viewer preferences appear as follows.

Viewer group	1	2
Program 1	10¢	10¢
Program 2		14¢

The monopolist can make the most money (a $550 profit) if it offers program 1 only. (It would earn only $450 if it produced both programs 1 and 2.) Although the monopolist responds to viewers' demands for preferred programs, the additional revenue generated by an additional program must cover the cost of the program and the revenues lost from other programs when viewers switch from them. In other words, the monopolist takes direct account of the lost profits on its other channels in its decision to broadcast an additional program.

Suppose that group 2 is still willing to pay 14¢ for program 2. Instead of having the two channels operated by a monopolist, however, they are operated by competing broadcasters. What programs will be offered viewers? One broadcaster will offer program 1 and receive a profit of $300, and the other will offer program 2 and receive a profit of $150. Note that without price competition, the second broadcaster would duplicate program 1, also charging viewers 10¢. In this case its profit would be $175, the first broadcaster's profit would also be $175, and a third broadcaster would enter to provide program 2. The likelihood of program duplication, however, is reduced by the first broadcaster's strong incentive to avoid duplication and entrants' recognition that the first broadcaster may reduce prices after they enter. In either case, competitors offer both programs, whereas the monopolist offers only program 1.

What do these two examples suggest? First, under pay television, a monopolist of all channels will respond to preferred choices because it can charge viewers according to their preference intensities. Second, in deciding to offer additional programs, the monopolist includes in its profit calculations the expected revenues lost on other channels. Third, under pay television (where channel capacity allows it), competitive broadcasters are likely to provide more programs than is the monopolist, because each competitor does not take direct account of the diminished revenues on other channels that result from its own choice of programs.

Another difference under monopoly and competition is important. Prices will probably be lower under competition, because competitors are forced to charge "competitive" prices.

Resource Allocation

The resources devoted to television and their allocation among program types differ under pay television and advertiser-supported television.[13] Under pay television, competitors engage in price competition and hence tend to avoid program duplication. The monopolist will produce preferred programs if viewers are willing to pay a profitable premium. The following examples will illustrate how viewer preferences would have to look for the revenue structure under pay television to resemble that under advertiser-supported television.

Assume that advertising revenue per viewing household for a half-hour program in prime time is 10¢. Assume for simplicity that viewers will watch only their first choices and that they will pay 10¢ to view their preferred programs. In this example revenues under pay television will be identical to revenues under prime-time, advertiser-supported television. Viewer preferences appear as follows.

Viewer group	1	2
Program 1	10¢	
Program 2		10¢

The likelihood that all viewers will be willing to pay 10¢ is remote. Some viewers may be willing to pay only 5¢; others will pay more. Advertiser-supported television registers no response to these differences, and the chances are great that advertising revenues will be substantially below what viewers are willing to pay. The resources allocated to pay television and the resulting program patterns will respond to these preference intensities.

Suppose all viewers are willing to pay 50¢. Then potential revenues under pay television are five times those under advertiser-supported television. This is not much different from Noll, Peck, and McGowan's (1973) estimate that viewers would be willing to pay up to seven times as much for network programs as the programs would generate in network advertising revenue. (To compare resources that would flow into program production under pay television and advertiser-supported television, one needs estimates of the differences between transactions and billing costs under the two structures.) Under pay television, program prices will be bid up and additional resources will flow into television. If there is adequate channel capacity, programs will be added at the margin, because minority programs that are not profitable at 10¢ per viewer will now be profitable at 50¢. If channel capacity is limited, additional programs will not be possible. In this case one might expect competitors to charge higher prices also but, in their competition for viewers, to produce or purchase more expensive, higher-quality programs.

Half-hour programs could be worth less than 10¢ to viewers. In this case pay television would allocate fewer resources to television than would advertiser-supported television.

Under pay television, not only total spending but spending on specific programs will be more responsive to viewer taste. In determining which programs to offer, broadcasters will weight the number of viewers who prefer a program by their preference intensities, as measured by the amounts they are willing to pay. To the extent that smaller groups are willing to pay a premium for their programs, then the profitability of these programs will reflect these viewers' desires.

A word of caution is needed. Pay television is often discussed in terms of the minority-taste programming that it may support. The existence of pay television does not necessarily imply that mass programs will disappear. Mass programs are still likely to be relatively profitable under pay television, either because viewers of these programs will be willing to pay relatively high prices or because the sheer number of potential viewers will make these programs profitable at low prices.

Pay television (particularly with unconstrained channels and a competitive structure) is more likely to allocate resources in television production efficiently than is advertiser-supported television. The reason is quite simple: revenues per viewer under pay television are more likely to reflect viewers' program preferences than are revenues per viewer under advertiser support (which reflect advertisers' values of viewers).

Program patterns under the two means of support differ in several ways. First, competitive program duplication is not nearly as likely to occur under pay television as under advertiser-supported television. Second, a monopolist controlling all channels in the market will have an incentive to produce preferred programs under pay television. The monopolist will have little incentive to do this under advertiser support, where viewers are worth equal amounts in advertising markets. Third, in pay-television markets with ample channels, the competitive channel structure, through competition for audiences, is likely to produce more programs (at lower prices) than is the monopolistic structure. In maximizing the joint profits of all channels, the monopolist takes direct account of how increasing its audience on one channel may decrease it on other channels. Finally, where the number of channels is ample, both the total resources devoted to television and the amounts allocated to individual programs are more likely to be economically efficient under pay television than under advertiser-supported television.

Summary

Economists have developed program choice models to understand the economic reasons for the program options offered viewers by the television industry. Steiner-type models dominated the analysis of program choice through the mid-1970s. The Beebe model, the capstone of the Steiner line of inquiry, identifies five factors that determine the range of television programs that are offered: (1) the structure of viewer preferences among program types; (2) the number of television channels; (3) the form of control of the channels—monopolist or competing broadcasters; (4) the means of support for programs—advertiser payments and/or viewer payments; and (5) the cost of a program.

If most viewers want the same types of programs and television is supported by advertiser payments, competing broadcasters are likely to offer highly similar programs targeted to this mass audience. From the perspective of social welfare, the number of programs designed to satisfy majority tastes will be excessive because competing channels will find it most profitable to carve up the majority-taste audience with close substitutes that do little to increase overall viewer satisfaction. Therefore, if the number of channels is small and most viewers have similar preferences among program types, viewers who highly value programs appealing to minority tastes may not be served. Competitive duplication leading to a narrow range of programming is less of a problem the less skewed (the more evenly distributed) viewer preferences are among program types, because the audience to be shared by duplicative channels is smaller. Even if preferences are highly skewed, minority tastes will still be served if the number of channels is large enough to exhaust the profits in competitive duplication, making programs for minority-taste audiences as profitable as majority-taste programs at the margin (assuming minority-taste audiences are large enough to cover program costs).

A monopolist of an advertiser-supported television system may do better than competitive broadcasters in avoiding duplicated programs and providing a wider range of program types, as long as viewers are not willing to accept common second-choice or third-choice programs in lieu of not viewing. A monopolist will never duplicate programs, because the expenditure on a second program of a given type will produce no net gain in audience. In general, a monopolist will add programs (and program types) as long as a new program increases the aggregate television audience by enough to cover the program's cost. Thus viewers' willingness to accept second and third choices among programs reduces the number of program types a monopolist will find it profitable to offer.

Advertisers' valuations of the audiences generated by programs are the sole determinant of which programs are provided by advertiser-supported broadcasters. Ignored are the intensities of viewers' preferences among programs. The problems of competitive duplication and of monopolists' failure to provide variety when viewers want a diversity of viewing options are ameliorated with pay support. Competitors will bid down the prices of their programs and/or bid up their production costs, either of which will reduce the number of firms that can profitably supply a single type of program. If television is supported by viewers, a competitive industry will generally provide more programs than will a monopolist, because the monopolist will count viewers and revenue diverted from other programs as a cost of a new program while competitive firms will not.

4 / Modern Models of Program Choice

The most serious limitation of Steiner-type models of program choice is their rather rude methodology for comparing the desirability of different outcomes. Merely tabulating the number of viewers receiving first choices, second choices, and so on does not take account of differences in the intensity of these preferences. The first choice of a viewer who has a hard time deciding what to watch receives the same weight as the first choice of a viewer who has a strong preference for one program over all others. The first significant break with the Steiner tradition was the Spence-Owen model (Spence and Owen, 1975, 1977; Wildman and Owen, 1985), which uses willingness-to-pay as a measure of preference intensity. Subsequent work in program choice has generally treated viewer preferences in this way, although the most recent research has employed models that are somewhat more specialized than Spence and Owen's. We begin with brief discussions of demand analysis and welfare analysis, which underlie the models presented in this chapter.[1]

Demand Analysis

Consider a simple case of three viewers (A, B, and C) who want to watch a single program, program 1. How much they want to see it is measured by how much they are willing to pay to watch the program if the alternative is not viewing. For any price greater than $7 and no more than $10, only one viewer will watch program 1. Two viewers will pay to watch the program if the price is between $4 and $7, and all three will pay to watch if the price is $4 or less (see Table 4.1 and Figure 4.1).

A profit-motivated seller will want to maximize the difference

Table 4.1 Three viewers' willingness-to-pay to see one program

	Viewers		
Program	A	B	C
1	$10	$7	$4

between revenue from sales and the cost of the program. Because the cost of the program does not vary with the number of purchasers, this means maximizing revenue. Maximum revenue will depend on the ability of the seller to discriminate among buyers in setting price. As we saw in Chapter 2, the seller that can charge different buyers different prices can exploit viewers' willingness-to-pay and earn larger revenues.

If the seller of program 1 is able to set different prices for each viewer, the seller will charge viewer A $10, viewer B $7, and viewer C $4. Total revenue will be $21. If the seller is unable to discriminate among viewers (for example, the seller may know the general distribution of willingness-to-pay, but be unable to determine the willingness-to-pay of any particular viewer), it will be forced to set a single price for all viewers. Here the profit-maximizing single price is $7. Two viewers paying $7 each will provide a total revenue of $14; three viewers paying $4 or a single viewer paying $10 will generate revenues of only $12 and $10, respectively.

Another way of looking at the same problem is to start at a price of $10 and then calculate the marginal changes in revenue associated with adding a viewer, given the price reduction required to do so. The marginal revenue generated by lowering the price from $10 to $7 is $4, the difference between $14 paid by two viewers paying $7 each and the $10 that viewer A would be willing to pay. The marginal revenue of adding viewer C to the audience by lowering the common price to $4 is a negative $2, because revenue drops by $2 to $12 if the price is lowered just enough to induce viewer C to buy. The profit-maximizing single price for program 1 is therefore $7, and the audience size is two. This illustrates a general principle: a seller facing a declining willingness-to-pay schedule and forced to set a single price will probably set the price high enough that some potential users of the product will not buy it.

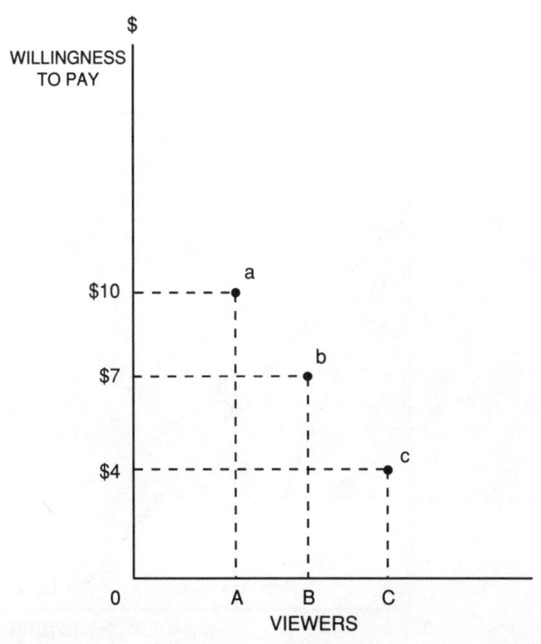

Figure 4.1 Three viewers' willingness-to-pay to see one program

The principles illustrated in this three-viewer example can be represented more generally. Assume that A, B, and C are only three of the many potential viewers of program 1 and that their particular willingness-to-pay are three points on a continuum of preference intensities of all viewers in the market. By filling in the gaps between the points represented by these three viewers, we can represent graphically the preference intensities (or willingness-to-pay) of all viewers in the market. This is the demand curve D_1 (in this case, a straight line) in Figure 4.2, which graphically represents consumer demand.[2] Note that D_1 falls from left to right, indicating that units sold increase as the price is lowered. If the demand curve is a continuous line, then the curve representing the marginal revenue associated with adding viewers incrementally must be continuous also. This is MR_1 in Figure 4.2. The profit-maximizing single price, P*, is the price at which marginal revenue goes to zero, or MR_1 hits the horizontal axis.

Demand analysis becomes more complicated when the demand for one or more other programs that are imperfect substitutes for

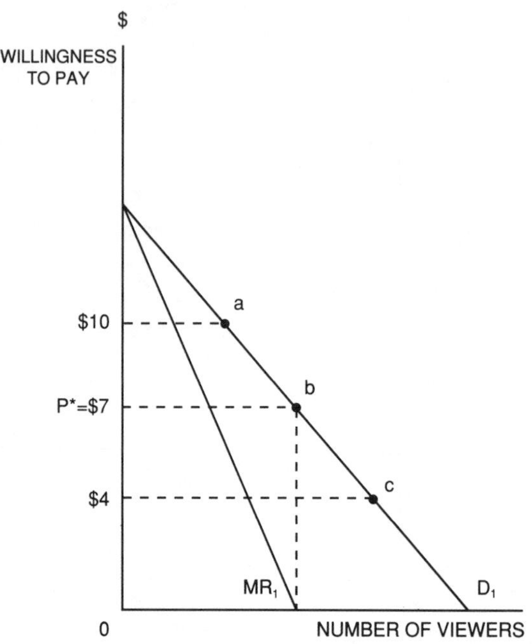

Figure 4.2 Demand for one program

the first program is considered. Programs (or products generally) are imperfect substitutes if consumers have preferences among them, but are willing to substitute one for another if the cost is not too great. Returning to the three-viewer example, assume that there is a second program, program 2, and that viewers would be willing to pay amounts indicated in Table 4.2 for it if it were offered on a stand-alone basis.

With two programs to choose from, a viewer will choose the program for which the difference between price and stand-alone value is largest. (We are assuming that a viewer will watch one of the two programs but not both. The analysis is more complicated if viewers can watch more than one program, but the results are basically the same as long as the value of any given program to a viewer declines the more programs are viewed.)

In other words, viewers will choose their programs to maximize the personal benefits of viewing after the price has been netted out. For example, if program 2 is free, then viewer A will be willing to pay no more than $3 to see program 1 and viewer 2 will pay no more than $2 to receive program 1. Viewer 3 will not watch program 1, even if it costs nothing. (In fact, it would be necessary to pay $6 to get C to watch program 1.) The change in the willingness-to-pay schedule for program 1 (program 1's demand curve) caused by the availability of program 2 at a zero price is illustrated in Figure 4.3. The substitute program shifts the demand curve down (or inward).

In general, the less the substitute costs, the larger the downward shift of the demand curve. Although the demand curve has shifted down, the benefits of the program to the viewer have not changed. Because the seller of program 1 will now have to sell the program

Table 4.2 Stand-alone willingness-to-pay schedule

	Viewers		
Program	A	B	C
1	$10	$7	$4
2	$7	$5	$10

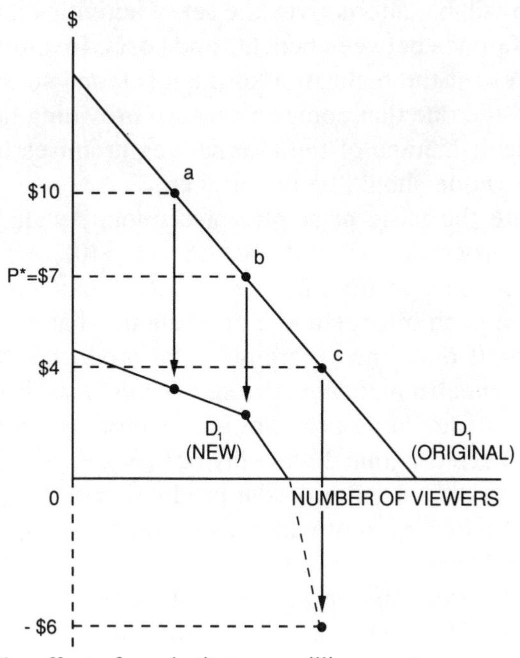

Figure 4.3 The effect of a substitute on willingness-to-pay

for less than $7, however, more of the benefits of viewing program 1 are retained by its viewers. (That is, the competition from program 2 has reduced the amount of program 1's value to viewers that the seller of program 1 can capture through the price charged.) For this reason, viewers generally benefit when competing channels reduce prices in an attempt to build audiences. A single seller controlling both channels would recapture some of the benefits transferred to viewers through competition by raising the prices of both programs.

Welfare Analysis

Economic welfare analysis assesses and compares the benefits and costs of alternative economic activities. Benefits are measured in terms of consumers' willingness-to-pay, and costs are the values (benefits contributed) of the resources employed in these activities in their best alternative uses. The use of willingness-to-pay as the measure of benefits makes economic welfare analysis a particular application of demand theory. Ideally, welfare analysis is used to select from available alternatives the set of activities that will maximize the difference between benefits and costs. Institutional limitations may prevent the realization of the ideal, and so the preferred alternative is the one that comes closest to providing the net benefits of the ideal. If none of the alternatives promises benefits that exceed costs, none should be undertaken.

To illustrate the basic principles of economic welfare analysis, assume that programs 1 and 2 cost $8 and $10, respectively, to produce. There are positive net benefits to producing either program, because each offers stand-alone benefits that exceed its production costs. If only one program can be produced (because, for example, the spectrum allocated to television is sufficient for only one channel), it should be program 1. Viewers are willing to pay a total of $22 to see program 2 and only $21 to see program 1, but $2 more in resources is used up in the production of program 2. Program 1 contributes $1 more to viewer welfare and therefore it should be produced.

Should both programs be produced if there are no barriers to doing so? No. If both programs are made available, viewer benefits can be increased to $27 with A and B watching program 1 and C watching program 2. Adding program 2 increases the willingness-to-pay measure of total viewer satisfaction by $6, but program 2 costs $10 to produce. Welfare would actually be reduced by $4, and therefore program 2 should not be produced.

The contribution of program 2 to total viewer benefits, $6, equals what viewers (in this case viewer C) are willing to pay to have program 2 made available if program 1 is offered at its initial price (in this case zero). The quality of these two measures is no coincidence. In general, the gain in aggregate viewer benefits from adding a new program to a given bundle of programs already available at predetermined prices is the sum of viewers' willingness-to-pay for the new program. When there are many such viewers, this gain can be calculated as the area under the demand curve for that program. If the cost of supplying additional viewers is zero, a key assumption of Steiner-type models and the Spence-Owen model, a program should be supplied as long as the total area under its demand curve is at least as large as the cost of producing the program.

The Spence-Owen Model

In the real world, television institutions fall short of the ideal of welfare maximization for a variety of reasons. Some programs that should be produced are not, and other programs are produced even though their costs exceed their benefits. The extent of these shortcomings depends on the structure of the industry and the manner in which financial support is generated. Which set of institutions best serves the objective of maximizing the difference between viewer benefits and program costs can be determined only by comparing one set of arrangements with another in terms of their departures from the ideal. Spence and Owen performed this operation for four sets of polar cases: monopoly and competition with pay (audience) support and monopoly and competition with advertiser support.

Pay Support

The Spence-Owen analysis of the pay-television industry makes the following assumptions:

1. The number of channels is unlimited, although all may not be used.

2. The costs of programs, determined exogenously, are fixed.
3. Competition takes place within a single program period.
4. No two programs are perfect substitutes, although the degree of substitutability may vary.
5. The cost of adding a new viewer to a program's audience is zero. In other words, delivery costs are independent of the size of the audience.

The Steiner models of program choice similarly allowed for unlimited channels and assumed competition within a single program period. But the Steiner models assume that programs of the same type are *perfect* substitutes, and therefore those models usually divide the audience for the type evenly among programs. The Spence-Owen assumption that all programs are somewhat differentiated seems to be more reasonable. Copyright laws alone demand some differences.

The Spence-Owen assumption, which is also employed in the Steiner analyses, that delivery costs do not vary with audience size, is usually true for over-the-air broadcasters but does not reflect the reality of positive hook-up costs for cable subscribers. Even in the case of cable, however, the assumption that new viewers can be added at no cost can be accepted as a convenient simplifying assumption that does not affect materially the nature of the conclusions.

Unlimited channels and differentiated products mean that the Spence-Owen model of competition among pay television services is a model of monopolistic competition in the spirit of Chamberlin (1956).[3] When each program is differentiated somewhat from all other programs, the demand curve for each program slopes downward. This means that a small increase in price does not drive all customers away, but only those who, at the new price, can realize equivalent or better net benefits from other available programs. These are viewers who see one or more other programs as relatively close substitutes for the program whose price is raised. Viewers with a stronger preference for that particular program will be more reluctant to drop it; thus as the price is raised, the audience for the program melts away only gradually.

Competition. We begin by developing the model as it is applied to a market of competitive pay services. Each service provides a single program. For this part of the analysis, the following assumptions are made.

1. New programs are introduced as long as it is profitable to do so (that is, as long as they generate revenues large enough to cover production costs).
2. The market for advertising is perfectly competitive.
3. In deciding whether to offer a program and how to price it, each programmer assumes that its competitors will respond by adjusting their prices to avoid any loss of audience.

The first of these assumptions is easily understood. Competitive firms do not supply products (or programs) on which they expect to lose money. The second assumption warrants additional comment. The Spence-Owen model is sufficiently general to allow programmers to charge viewers and to sell commercial time to advertisers that want to reach these viewers. This is the situation with most basic cable channels. Cable subscribers pay a fee to receive the program services in the basic package, but most of these services have commercials. Advertising time is priced at exactly its value to advertisers. This means that the price of advertising time, or exposure to audience members, is equal to its contribution to economic welfare. In our analysis we will assume first that all channels are supported by viewer payments alone and then that they are supported by advertiser payments and viewer payments.

The third assumption is known as the Cournot assumption. To model competition among economic agents, it is necessary to specify how each agent expects its competitors to respond to its own actions. Specifications of anticipated reactions by competitors are referred to as conjectural variations. Spence and Owen used the analytically convenient Cournot conjectures in the model, but other reasonable assumptions concerning conjectural variations can be used without changing the primary qualitative results.

One implication of profit maximization with downward-sloping demand curves is that prices will be set too high to maximize the viewer benefits associated with the provision of each program. The profit-maximizing price and audience size for a representative program in a competitive pay-television industry are illustrated by P^* and V^* in Figure 4.4. Profit maximization dictates that price be

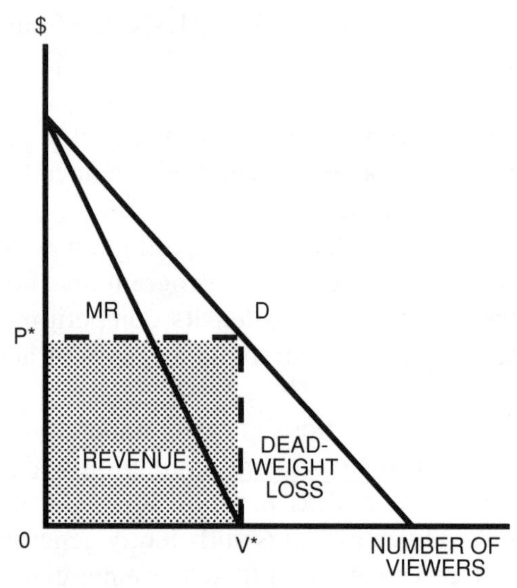

Figure 4.4 Profit maximization

set at the level at which the marginal viewer's contribution to revenue (marginal revenue) is equal to the cost of adding the marginal viewer to the audience (marginal cost). Marginal cost is zero. Therefore, the price is set to generate the size of audience at which the marginal revenue curve intersects the horizontal axis. Welfare maximization requires that price be set such that no viewers who value the program more than the cost of providing it are excluded. The benefits from viewing that these viewers would have received if the program were available free is a welfare loss, sometimes called a deadweight loss, associated with this type of pricing strategy.

Figure 4.4 also illustrates another failing of monopolistic competition relative to the welfare ideal: a new program should be added as long as the value of the program to viewers, which is the area under the program's demand curve, exceeds the cost of producing it. Yet competitive programmers will supply programs only if the revenues generated are at least as great as program production costs. Revenue is price multiplied by the number of viewers, the shaded rectangle in Figure 4.4. This falls short of viewers' valuations of the program by the sum of the areas of the lower right-hand triangle of deadweight loss, which is the valuations of viewers who are not willing to pay the price charged, and the upper triangle, which is the amount by which the valuations of viewers who do subscribe to the program exceed the price they pay.

Only programs that can generate revenues large enough to cover production costs will be supplied by a competitive pay-television industry. For this reason even programs that generate viewer benefits greater than costs will not be supplied if their revenues are insufficient. A competitive pay-television industry is biased "*against programs that have demands such that revenues capture a small fraction of the gross benefits*" (Spence and Owen, 1975, p. 151, emphasis in original). If two programs cost the same and promise the same total benefits to viewers, a pay channel will provide the program for which it can extract the greatest revenues. Among marginal programs (programs that just break even), there are likely to be programs that generate just enough revenue to cover costs but, because of the shapes of their demand curves, contribute less to the net benefits of viewing (viewer benefits minus production costs) than do some programs that are not provided. Programs whose revenues are low in comparison with viewer benefits are programs for which the benefits of viewing are highly concentrated in a relatively small fraction of the viewers who would be willing to watch it at a zero price. This condition is reflected in demand curves that are convex to the origin (bowed in). Sellers are able to capture a larger fraction of viewer benefits if demand curves are concave to the origin (bowed out). Linear demand curves, as in Figure 4.4, represent an intermediate case. Convex and concave demand curves, and the portion of viewer benefits that can be captured as revenue under each, are illustrated in Figures 4.5 and 4.6.

In addition to being biased against programs that are highly valued by a few viewers, competitive pay-television services are baised against expensive programs. If two programs provide the same benefits once program costs have been netted out, a pay channel will provide the less expensive of the two. In some circumstances, less expensive programs will be substituted for more expensive programs that would have provided greater net benefits (see Figures 4.7 and 4.8). The gray rectangle in Figure 4.7 is the revenue such a program can generate if profits are maximized.

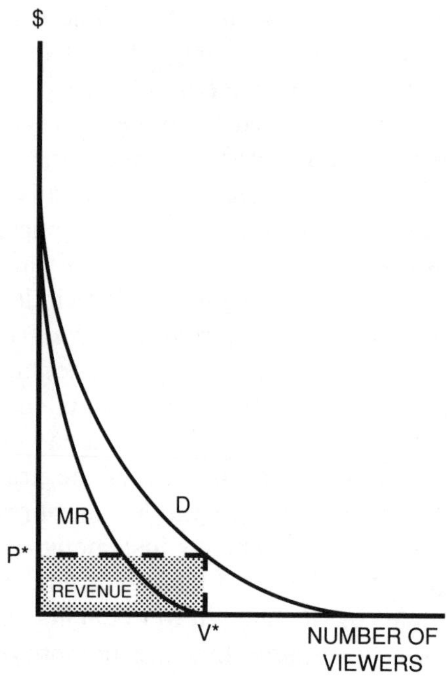

Figure 4.5 Convex demand curve

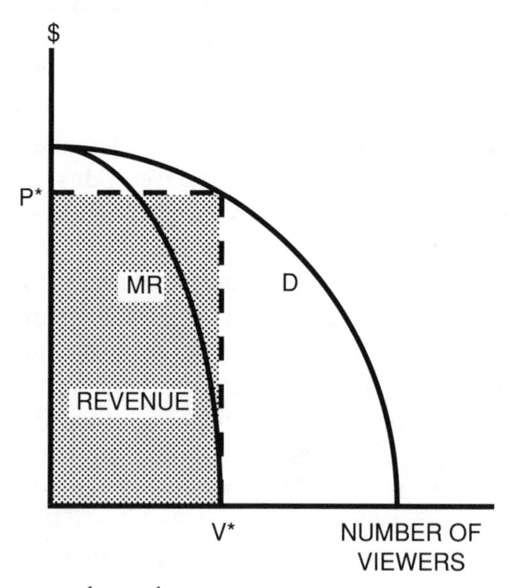

Figure 4.6 Concave demand curve

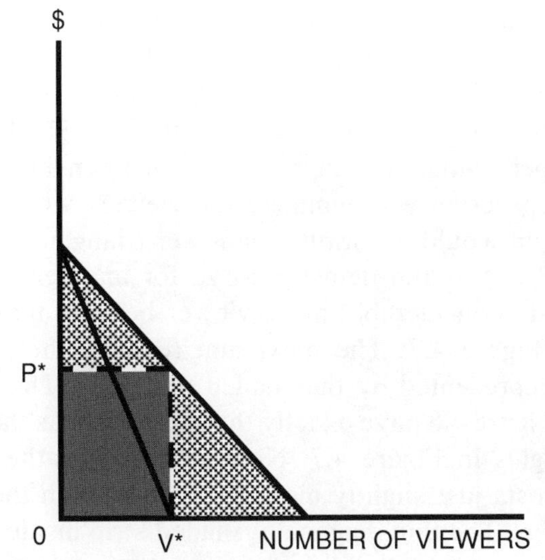

Figure 4.7 Demand and revenue for less expensive program

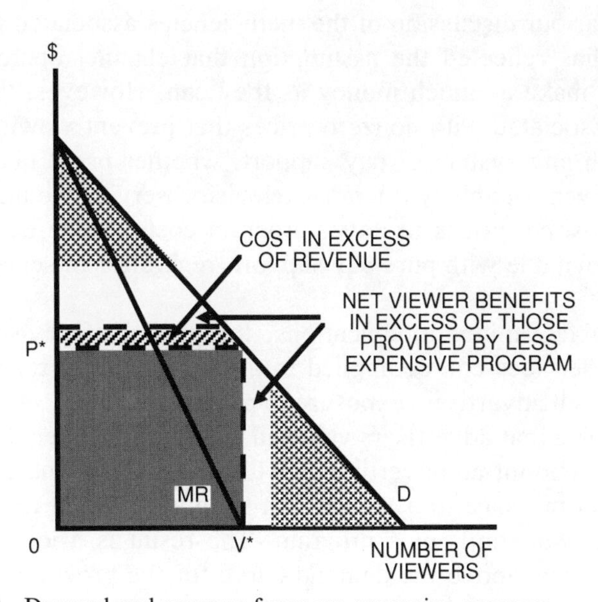

Figure 4.8 Demand and revenue for more expensive program

Assume that the cost of the program is just equal to the size of this rectangle. This program will be produced by a competitive television industry. The potential viewer benefits of the program in excess of the cost of the program are the two shaded triangles. Only the upper triangle will be realized as net benefits, because the price of the program will eliminate the viewers whose enjoyment of the program would constitute the lower triangle.

Figure 4.8 shows the demand curve for another program that would provide considerably more viewer benefits than would the program in Figure 4.7. The maximum revenue the program can generate is represented by the shaded rectangle. The two shaded triangles in Figure 4.8 have exactly the same areas as the net viewer benefit triangles in Figure 4.7. Now assume that the program in Figure 4.8 costs just slightly more to produce than the revenue it would bring in. The thin diagonally shaded strip inside the demand curve just above the revenue rectangle represents the amount by which cost would exceed revenue. A competitive television industry with pay support will not produce this program. Yet this program would provide net benefits that exceed the net benefits of the first program by the sum of the unshaded areas under its demand curve.

Thus far our discussion of the inefficiencies associated with pay support has reflected the assumption that channel operators are trying to make as much money as they can. However, the inefficiency associated with nonzero prices that prevent viewing would arise with any system of pay support, whether profit maximizing or not. Even a publicly operated television service would have to charge positive prices to cover program costs. Some deadweight loss is inevitable with pure pay support, regardless of sellers' motivations.[4]

With a few minor modifications, this model of pay-supported program services can be applied to a television industry in which channels sell advertisers exposure to paying audiences. Assume for convenience that advertisers value all audience members the same. Then the amount an advertiser is willing to pay for the chance to present its message to a viewer can simply be added to audience members' valuations of a program. The result is a joint benefits curve that lies above the demand curve for the program by z, the amount (assumed to be the same for all advertisers) that advertisers' value exposure to an audience member. The joint benefits curve gives the sum of what advertisers and viewers are willing to pay for the program as a function of the size of its audience. Therefore, programmers will maximize profits with respect to the joint benefits curve as if it were a demand curve. A joint benefits curve, labeled JB, and its relationship to the demand curve for a program are illustrated in Figure 4.9. Corresponding to the traditional marginal revenue curve, there will be a curve that reflects the marginal contribution of an additional viewer to the sum of viewer benefits and advertiser benefits. This is M_{jb}. The profit-maximizing channel operator will set the price to generate the audience, V_a^*, for which the sum of advertisers' payments and viewers' payments is maximized, which is determined by the point at which M_{jb} equals zero. Note that V_a^* is greater than V^*, the profit-maximizing audience for the program if viewer payments are the only source of revenue.

P^* and P_a^* are the prices that a profit-maximizing channel operator would charge without and with advertiser support, respectively.

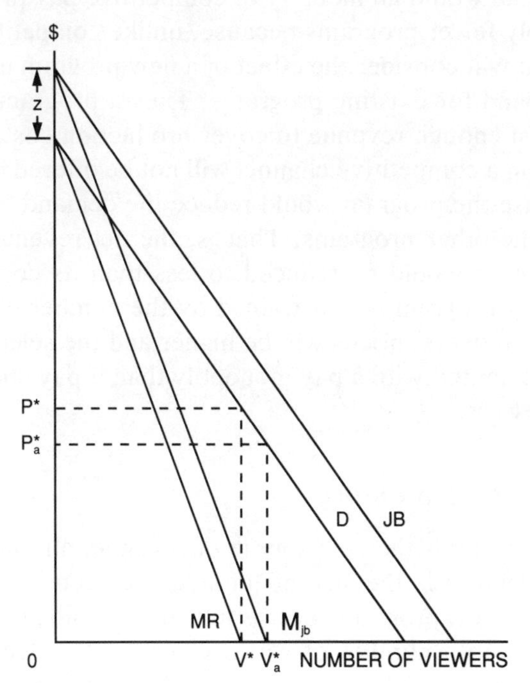

Figure 4.9 Joint benefits

Viewers get a better deal when advertisers contribute to the support of programs. Programmers profit more from the marginal viewers when advertisers will pay to reach them, and therefore programmers lower their prices to increase the size of the audience. The basic conclusions of our welfare analysis of a pay-only television industry are unchanged by the addition of advertiser support. All of the results for simple demand curves apply equally to joint benefit curves that include the benefits of audience members to advertisers, because the calculus for maximizing profits is unchanged.[5]

Monopoly. A monopoly supplier of pay programs will exhibit all of the biases and inefficiencies just discussed for competitive pay programmers. It will set prices high enough to exclude some viewers who otherwise would enjoy watching its programs. It will fail to offer programs that would provide viewer benefits that exceed costs but cannot generate enough revenues to cover those costs. It will tend to select programs for which it is possible to capture a large fraction of viewers' willingness-to-pay. In addition, a monopolist will tend to charge higher prices and supply fewer programs than would an industry of competitive pay programmers. It will supply fewer programs because, unlike competitive channel operators, it will consider the effect of a new program it might offer on the demand for existing programs. Thus a program that would generate just enough revenue to cover production costs and would be offered on a competitive channel will not be offered by a monopolist, because the program would reduce the demand for and revenues of some other programs. That is, the net revenue generated by the program would be reduced to less than its cost. Since the price for any program is constrained by the number of substitutes available to viewers, prices will be higher and the selection of programs more limited with a pay monopoly than if pay channels compete for viewers.[6]

Advertiser Support

When advertising is the sole source of revenue, all that matters in program selection is the size of the audience generated at a zero price. As a result, program choices fall short of the ideal of welfare maximization in a number of ways. Consider the demand curves for programs 1 and 2, shown in Figure 4.10. Even though the viewer

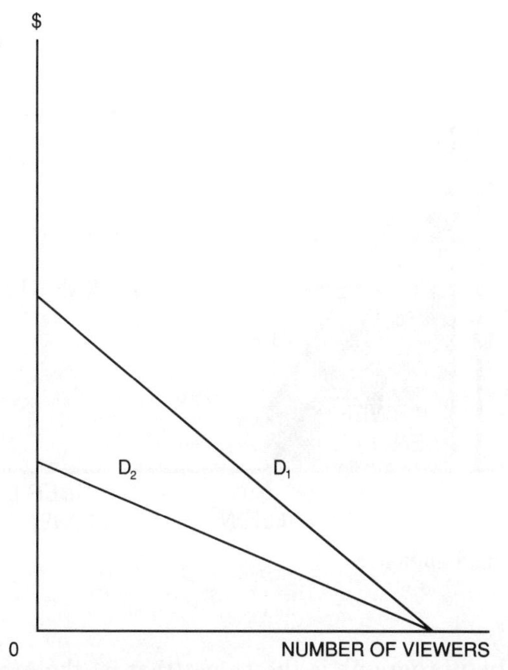

Figure 4.10 Demand for two programs

benefits of program 1 are much greater than the viewer benefits of program 2, an ad-supported programmer will not be more inclined to show program 1.

The bias of pay channels against programs for which viewer benefits are concentrated in a small portion of the audience is even greater for advertiser-supported channels. The preferences of the few viewers who value a program highly may be overshadowed by the lower willingness-to-pay of a mass of viewers less excited by the program. The higher willingness-to-pay of the few, however, will be reflected, if inadequately, in the aggregate demand curve for the program. Strong viewer preferences cannot be expressed at all with advertiser support. Therefore, an advertiser-supported industry is less likely than an industry of pay channels to produce programs highly valued by those viewers who do not share majority tastes.

There is a related bias in favor of programs with large audiences. In Figures 4.11 and 4.12, the sum of viewer and advertiser valua-

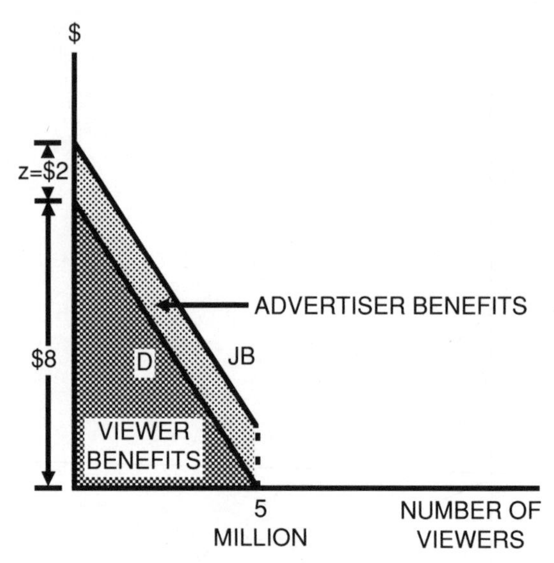

Figure 4.11 Small audience

tions of the two programs is the same (that is, the areas under the joint benefit curves are the same). But because more of the benefits are advertiser benefits for the program in Figure 4.12 than for the program in Figure 4.11, a competitive ad-supported channel choosing between the two programs will choose the one with the larger audience. This is a bias in favor of advertiser interests and against viewer interests.

The bias against programs with viewer benefits concentrated in a small portion of the audience and the bias against programs with small audiences are two aspects of a more general bias against programs with steep demand curves (or inelastic demands). Advertiser support also biases program selection against costly programs. The less costly programs generally will be provided when programmers must choose between high-cost programs and low-cost programs that promise the same net benefits. This can be seen by considering the programs whose demand curves are depicted in Figure 4.10. With audiences of the same size, the two programs would generate the same advertising revenue. Suppose the cost of program 1 (the interior demand curve) is just equal to the value

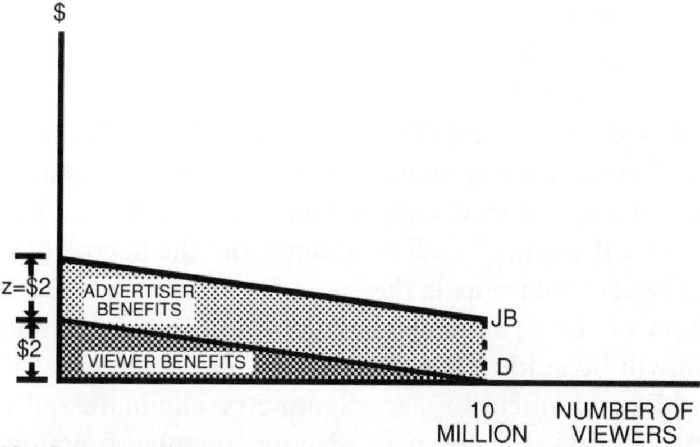

Figure 4.12 Large audience

of the audience to advertisers. The viewer benefits of program 2 exceed the viewer benefits of program 1 by the area between D_1 and D_2. Suppose also that it costs more to produce program 2 than to produce program 1, and the difference in costs is less than the difference in the viewer benefits of the two programs. A programmer choosing between them will select program 1, even though program 2 provides greater net benefits.

Pay television and advertiser-supported television are both biased against minority, or special interest, tastes and expensive programs, but these biases are more pronounced with advertiser support. From the point of view of biases, pay television is preferable to advertiser support. However, the zero price for advertiser-supported programs means that no viewer who enjoys a program will be deterred by its price. Therefore pricing is more efficient with advertiser support. Additional analysis is required to determine which type of industry produces the greatest net benefits overall.

Welfare Comparisons of Industry Structures

A competitive pay industry almost always is preferable to a pay monopoly controlling all channels.[7] But does competitive pay television, competition with advertiser support, or an advertiser-supported monopoly of all channels provide the greatest net benefits (or total surplus) of television viewing? There is no obvious choice among these industry structures. Each may be the preferred choice under the right circumstances.

Spence and Owen simplified the calculation of welfare benefits for the different industry structures by assuming all channels to be symmetrically positioned with respect to each other. This means that costs are the same for all programs, and the degree of substitutability between programs is the same for all program pairings. The implication of the symmetry assumption is that in equilibrium all channels will be sold at the same price, and audiences for all programs will be identical in size. Symmetry eliminates all biases in program selection, but the gain is more tractable functional forms for welfare comparisons.

Comparison of the welfare benefits of the alternative industry structures is facilitated by locating the equilibrium for each industry structure on a graph, with the number of programs on one axis and the average audience size (viewers per program) on the other. In Figure 4.13 the competitive pay equilibrium is indicated by E; the pay monopoly solution by M; the competitive equilibrium with advertiser support by CA; and the solution for an advertiser-supported television monopoly is MA. Finally, O is the welfare optimum subject to the constraint that programs are provided free

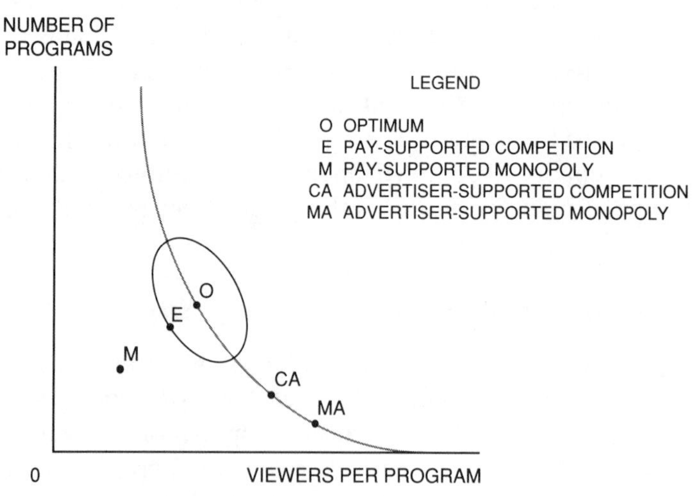

Figure 4.13 Comparison of welfare benefits

to viewers. As discussed in note 5, welfare could be improved if viewers were paid as much to watch programs as the value of the average audience member to advertisers. Such a pricing scheme is probably impractical, however, so the welfare optimum subject to the pricing constraint (the second best optimum) is that point at which total surplus is maximized with prices set to zero.

It is fairly simple to determine the relative positions of O, E, and M. Recall that competitive pay channels will not be able to collect as much in revenue as the viewer benefits their programs provide, so some programs that would contribute to total surplus will not be supplied. A competitive pay-television industry will therefore provide fewer than the optimum number of programs. Because these programs will be priced too high for welfare efficiency, some viewers will be excluded who could benefit from watching the programs supplied. For this reason, audiences will be smaller than they should be for welfare maximization. E will therefore lie below and to the left of O. A pay monopolist will provide fewer programs and charge higher prices than will a competitive pay industry; so M lies below and to the left of E.[8] Net viewer benefits are reduced by the movement from O to E and by the shift from E to M. Therefore, E must be superior to M in terms of welfare performance.

The positioning of CA and MA relative to O and E is more difficult. It depends primarily on the degree of substitutability among programs and whether the value of the marginal program to viewers exceeds the value of the marginal program's audience to advertisers. We have drawn through E a line connecting all combinations of audience size and number of programs that provide exactly the same total surplus as does E, where total surplus is the sum of total net benefits to viewers and to channel programmers. Clearly, O must lie within the isototal surplus oval through E, and M must be outside this boundary.

The line through O is the schedule of all combinations of audience size and number of programs that is consistent with offering programs at a zero price. This schedule slopes downward from left to right because decreasing the number of programs will lead to an increase in the average audience per program, as audience members reallocate themselves among the remaining programs. CA and MA must lie along this schedule through O. An advertiser-supported monopolist will consider the fact that new programs will cut into the audiences for existing programs; competitive advertiser-supported channels will not. For this reason the competitive industry will offer more programs, so MA must lie below and to the right of CA. We have located CA and MA outside the isototal surplus line through E. In the situation depicted, the competitive pay industry provides greater net benefits than does a competitive advertiser-supported television industry or a monopoly of advertiser-supported channels. However, CA and/or MA could lie within the isototal surplus line through E (thus providing greater total benefits than provided by a competitive pay industry).

Which of the alternative equilibria represented by E, CA, and MA will generate the greatest total surplus depends on viewers' perceptions of the substitutability of one program for another. To see this we take as an arbitrary starting point the positions for E, CA, MA, and O indicated in Figure 4.14 and track their movements relative to each other if programs become increasingly close substitutes. The welfare justification for having more than a single program is that the benefits to viewers of increased differentiation outweigh production costs. The stronger viewers' preferences are for differentiated fare, the more programs are required to maximize welfare. If tastes change and programs are seen as increasingly close substitutes, the benefits of differentiation decline and the number of programs required to maximize total surplus declines.

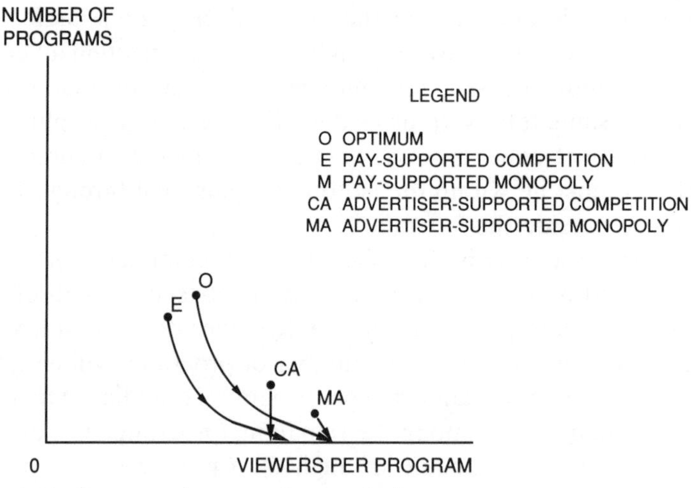

Figure 4.14 Programs become closer substitutes

Some programs then have to be eliminated, and the viewers of those programs choose from among the remaining programs, which increases their audiences. Therefore O moves down and to the right as programs become closer substitutes, as indicated by the arrow emanating from O in Figure 4.14.

Viewers' willingness-to-pay for program differentiation will also decline if programs became closer substitutes. For a competitive pay-television industry, this means a reduction in viewer payments overall and fewer programs that can be supported in equilibrium. Again, viewers are then redistributed among the remaining programs, whose audiences grow. Thus E would move in the same general direction as O if substitutability among programs were to increase.

Increased substitutability will reduce the equilibrium number of programs in a competitive, advertiser-supported television industry, because new programs will not bring in as many new viewers as before. The average number of viewers per program will remain the same, however, regardless of substitutability, because competitive entry will ensure that the advertising revenues for each program will just cover program costs. If the cost of a program and the value of a viewer to advertisers are unchanged, the audience per program must be constant also. CA will follow the vertical path downward indicated in Figure 4.14, and the CA path may intersect the path for O.[9] It is possible that CA could lie very close to the optimum at O.

An advertiser-supported monopolist will reduce the number of programs offered if substitutability among programs increases, because the contribution of the marginal program to the aggregate audience and advertising revenue will decline. Audiences for remaining programs will grow. MA will move to the right and downward as indicated, but the rate at which MA will move right will be less than for O. The paths of MA and O must converge when all programs are perfect substitutes. With perfect substitutability viewers do not benefit from having more than one program, so welfare is maximized with a single program. An advertiser-supported monopolist will provide just one program, because additional programs would only fragment the existing audience while multiplying program expenses. Therefore, if all programs are perfect substitutes (or equivalently, if all viewers have identical preferences among programs), it is best to organize the television industry as an advertiser-supported, single-channel monopoly.

In sum, a competitive pay-television industry, a competitive advertiser-supported industry, or an advertiser-supported monopoly may be the preferred industry structure in terms of welfare benefits. Which is best depends on the extent to which viewers perceive different programs to be good substitutes for each other. A competitive pay industry is always preferred to a pay-television monopoly. If programs are perfect substitutes, an advertiser-supported monopoly performs best. As programs become less acceptable substitutes for each other, the welfare preferred industry structure shifts first to ad-supported competition and then to competition with pay channels.

Spence and Owen (1975, 1977) argued that the television industry's contribution to economic welfare would be increased by a movement from what was then primarily an advertiser-supported equilibrium dominated by over-the-air broadcasters to a competitive pay-television industry. Noll, Peck, and McGowan (1973) had estimated the potential additions to viewer benefits to be $4.7 billion for a fourth broadcast network and $3.8 billion for a fifth network. At the same time, the average annual costs of programming and running CBS, NBC, and ABC were approximately $800 million apiece. Although the incremental viewer benefits apparently exceeded the costs of adding new networks by a considerable amount, advertising revenues would have been insufficient. According to Park (1973), a fourth broadcast network would not have been profitable. It was not until the late 1980s that we saw in Fox what may become a fourth full-fledged, over-the-air network. The explosive growth in cable penetration and VCR ownership since the Spence-Owen model was published in 1975 suggests that their assessment of viewers' willingness-to-pay for additional television channels was correct.

The Wildman-Owen Model

When Spence and Owen compared the options of pay and advertiser support for monopoly and competitive industry structures in the 1970s, television was primarily an off-the-air, ad-supported medium. The industry in the 1990s is a hybrid of the industry structures they considered. Advertiser-supported, over-the-air broadcasters compete with basic cable channels supported by ads and viewer payments; and both compete with pay cable channels such as HBO and Showtime, pay-per-view cable services, and video cassettes, all of which depend almost entirely on viewer payments.[10] Wildman and Owen (1985) extended the Spence-Owen model to examine television competition when pay and ad-supported services compete with each other. They also modified the model to make the amount of program time devoted to commercials a decision variable for programmers.

The latter modification was made necessary by the Justice Department's successful challenge to the code of the National Association of Broadcasters that had capped the amount of commercial time for member stations (see Chapter 5). Since 1981, when the code's commercial time restrictions were dropped, broadcasters have had to decide for themselves how much program time to allocate to commercials. In making this decision they must consider the effects of commercials on viewers' choices among channels. We assume that, other things being equal, viewers prefer programs without commercials to programs with commercials.[11] Readers value advertisements in newspapers as a source of information that is helpful to shopping. Yet there is abundant evidence that television viewers generally prefer programs without commercials. Viewers pay a premium for the pay cable channels that are commercial free; many viewers use remote controls to hop among channels during commercial breaks (behavior that is sometimes called "grazing") and to erase commercials on programs they have recorded. Independent stations sometimes promote programs with limited commercial interruptions. The difference in attitudes toward print and broadcast advertising may reflect the fact that readers can control the timing and duration of their exposure to advertisements while television viewers cannot, unless they use their remote controls, and even then they cannot avoid the interruptions commercials cause in the flow of a program.

The following analysis does not depend on the assumption that viewers dislike all television commercials. It is sufficient that viewers dislike commercials at the margin. Even if there is some range over which increasing commercial time does not take away viewers, a profit-maximizing broadcaster will always increase its commercial time to the point at which a marginal increase begins to have a negative effect on audience size.

If viewers do not like commercials, then commercial time may be treated as a nonmonetary price that viewers pay to see programs supported by advertising. As with monetary prices, we can draw demand curves relating the size of a program's audience to the "price" viewers pay in terms of the amount of advertising time inserted into the program (see Figure 4.15). As with a demand curve reflecting a monetary price, the number of viewers declines continuously as commercial time increases, because viewers differ in their tolerance for commercials in the program. Those most sensitive to commercials leave the audience first, followed by viewers increasingly more tolerant of advertising as the amount of advertising increases. Last to leave are viewers who are either relatively unbothered by commercial interruptions or viewers who value the program content highly and cannot find good substitutes in other programs.[12]

If the advertising market is competitive, the price advertisers pay for a minute's exposure to viewers will be constant. In this case, the profit-maximizing amount of commercial time for a program supported entirely by commercial time sales is the same as the

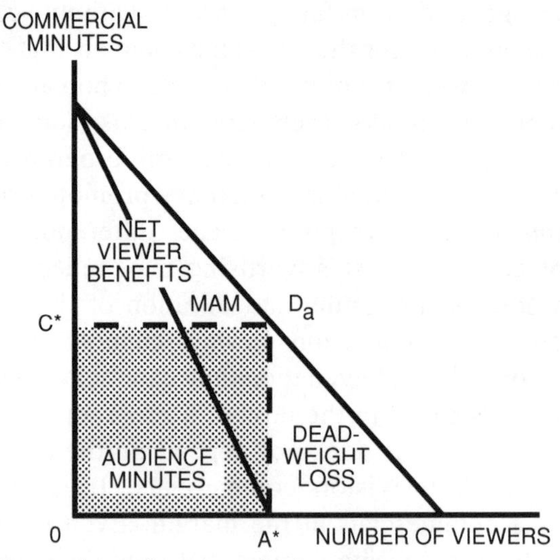

Figure 4.15 Demand as a function of advertising

amount of commercial time that maximizes the number of audience minutes sold, which is the number of advertising minutes times the size of the audience. For a given amount of advertising time in a program, audience minutes is the area of the rectangle under the advertising demand curve. The area is determined by drawing a horizontal line from the amount of advertising time on the vertical axis to the demand curve and then drawing a vertical line from that point on the demand curve down to the horizontal axis. The profit-maximizing level of advertising time is the amount that maximizes the area within this rectangle.

The parallel to maximizing profits under a standard (price) demand curve is straightforward. The profit-maximizing price produces the largest revenue rectangle (price times audience) under the standard demand curve. Corresponding to the marginal revenue curve under a standard (price) demand curve is a marginal audience minutes curve (MAM) under an advertising demand curve (D_a). Profits are maximized where marginal audience minutes are zero, or where the marginal audience minutes curve intersects the horizontal axis. The profit-maximizing audience and amount of commercial time are A* and C* in Figure 4.15. Note that, with the exception of the labeling, this looks exactly like profit maximization under a traditional demand curve.[13]

The profit-maximizing calculus is more complex for programs supported by viewer fees and advertising sales. In setting the amount of advertising time, the programmer must consider the effect of commercials on viewers' willingness-to-pay; in setting price, the programmer must consider the effect on the number of viewer minutes that can be sold to advertisers.

Figure 4.16 illustrates these trade-offs in terms of the total revenue generated for a channel by a representative viewer. The two fundamental relationships that must be considered by the channel programmer are the viewer willingness-to-pay schedule, W, and R_a, which is advertising revenue per subscriber. P_{max}, the upper left end of the willingness-to-pay schedule, is the amount the viewer is willing to pay to add the channel as a viewing option if the program carries no advertising.[14] The lower right end of the willingness-to-pay schedule, A_{max}, is the amount of advertising time required to drive the viewer's value of the program to zero. The willingness-to-

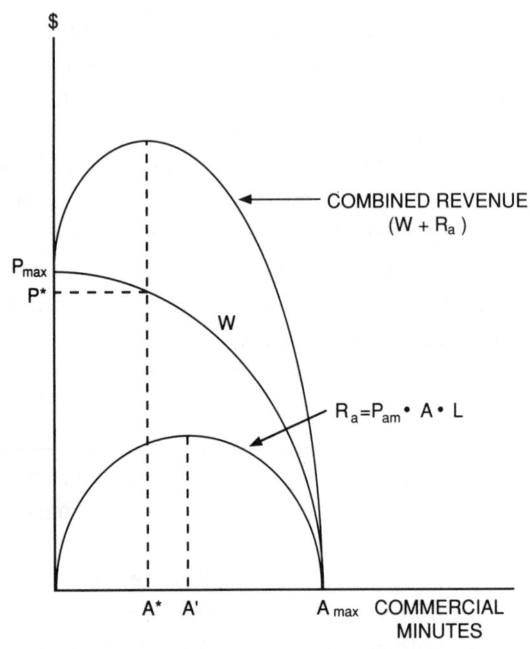

Figure 4.16 Combined revenues from a representative viewer

pay schedule falls from left to right because commercials detract from a viewer's enjoyment of a program; therefore, the amount a viewer is willing to pay for a channel will fall if the channel increases the amount of advertising time it sells.

For this analysis, it is important only that the willingness-to-pay schedule have a negative slope. As drawn, willingness-to-pay falls slowly at first, but faster as more commercial time is added. This seems plausible, because a small amount of advertising time can be placed at the beginning and at the end of a program and perhaps at natural breaks within a program without detracting too much from program continuity and without cutting too severely into the time required to develop a story. At some point, however, the loss of additional time to commercials constrains what can be accomplished artistically, and commercial interruptions are likely to become irritating to viewers.

Advertising revenue per subscriber, R_a, is the product of three terms: P_{am}, the amount an advertiser is willing to pay for a minute of commercial exposure to a representative viewer; A, the number of advertising minutes in a program; and L, the likelihood, or probability, that a given subscriber will be in the audience for any randomly selected program on the channel. R_a necessarily is zero if there is no advertising. So R_a rises initially as commercial time increases, and is concave from below. That is, R_a increases initially as A increases, but the rate at which ad revenue per subscriber can be increased by increasing the amount of advertising becomes smaller as A becomes larger. Eventually, increasing commercial time will lower ad revenue per subscriber. At A_{max} of commercial time, the viewer will not watch the channel at all. As A increases, L becomes smaller and smaller, eventually reaching zero at A_{max}. Ad revenue per subscriber will be maximized for some value of A smaller than A_{max}, which we label A′.

Combined revenue is the sum of willingness-to-pay and advertising revenue per subscriber. For the viewer in Figure 4.16, advertising revenue rises more rapidly than willingness-to-pay falls at low levels of advertising. Therefore combined revenue rises to a maximum that is greater than willingness-to-pay without advertising. Because the willingness-to-pay component of combined revenue always falls, combined revenue must peak at a lower level of advertising than does the advertising component of combined revenue by itself. This is indicated by A*. Because some advertising is provided, the profit-maximizing viewer payment, P*, must be less than P_{max}. Therefore, if the viewer depicted truly is representative and the programmer sells advertising time and charges viewers, the subscription charge will be lower than the price would be if the program were supported by viewer payments only, and less commercial time will be sold than would be the case if the channel were entirely ad supported.

Most cable channels, and the channels that account for most of the cable audience, are part of basic packages supported by both advertising revenues and viewer charges. This suggests that the relationships illustrated in Figure 4.16 are common. Some pay-only cable services exist because there are viewers for whom willingness-to-pay always falls more rapidly than R_a rises and/or there are certain program types for which this is the case for a large number of viewers. Basic and pay tiers of program services are created so that cable operators can more effectively exploit these distinct demands.

Regulation and current technology have forced broadcasters to support their services with advertising alone. From Figure 4.16 we can see that their inability to cash in on viewers' willingness-to-pay for their services is a disadvantage relative to cable services that can command direct support from viewers. However, this does not mean that "free" television is doomed in the face of ever-growing competition from subscriber-supported services. Viewers differ in their willingness-to-pay, and the selection of a profit-maximizing price necessarily excludes some viewers from the audiences of pay services. For example, about 40 percent of homes passed by cable do not subscribe. Even in the absence of the FCC rule requiring cable systems to carry all local broadcast signals, cable operators have found it profitable to carry most local broadcast stations. This means that broadcasters compete for shares of that part of the audience subscribing to cable, while cable services are excluded from households unwilling or unable to pay the price of cable service. Measured as a fraction of the universe of cable subscribers, broadcasters compete for an audience that is approximately 80 percent larger than the audience subscribing to cable services. Even if all homes had access to cable, at current subscription rates the potential audience available to broadcasters would still be two-thirds larger than the audience available to cable-only services. Furthermore, there are fewer competitors for the broadcast-only portion of the television audience. Therefore, it is not surprising that, mirroring the actual relationship between the audiences of broadcast networks and cable networks, broadcasters had much larger audiences in Wildman and Owen's simulation study of competition between advertiser-supported broadcast services and pay services.

How are the Spence-Owen welfare conclusions affected by the Wildman-Owen extension of the original model? Spence and Owen concluded that economic efficiency probably would be improved by a shift from the predominantly advertiser-supported television industry of the mid-1970s to an industry supported more by viewer payments. Recognizing that advertising constitutes a nonmonetary cost to viewers reinforces that conclusion, because Spence and Owen did not consider the effect of advertising on viewer welfare in comparing equilibria with different types of support. Given that viewers will differ in their willingness-to-pay for programming and their tolerance for commercials, aggregate viewer welfare probably would be maximized if there were a mixture of services, including pure pay services, services supported by advertising only, and program services supported by advertising and viewer payments. Wildman and Owen (1985) suggest that advertiser-supported services are likely to be underrepresented in a competitive equilibrium relative to their contributions to economic welfare. But the need to employ restrictive assumptions in simulation models means that caution should be exercised in generalizing from the study results.

Multichannel Programming

The FCC's duopoly rule makes it unlawful for a single agent to program more than one broadcast television channel in any television market, and therefore multichannel operations are not a feature of traditional over-the-air television service. Broadcasters compete with multichannel cable operators in most markets, however, and are also encountering new multichannel competitors in the form of multipoint multichannel distribution services, sometimes called "wireless cable" operations. Announcements of direct broadcast satellite services suggest that yet another multichannel service may soon vie for viewers' attention and pocketbooks. The persistence of grandfathered AM-FM radio combinations in many markets suggests that multichannel operations also could be a stable part of over-the-air television service if federal law did not prohibit it.

We have already examined certain aspects of multichannel operations in our discussions of the Steiner-type and Spence-Owen analyses of a monopolist of all television channels. Spence and Owen show that a pay-television monopolist would supply fewer channels of programming and would price them higher than would an industry of competitive single-channel operators. The monopolist sets a higher price because viewers lost to one channel owing to an increase in its price will be recaptured, at least in part, by increased demand for other channels controlled by the monopolist. If advertising is treated as a price to viewers, as in the Wildman and Owen analysis, then a monopolist will also sell more advertising time than would competitive channels, again because some of the viewers lost to a channel if its advertising time is increased will show up as viewers of other channels controlled by the monopolist.

The same principles applied by a multichannel monopolist to

setting prices and the amount of advertising time will also be applied, but with less force, by any multichannel program service, as long as viewers see the channels offered as substitutable for one another. Suppose a multichannel operator believes that some of the viewers lost because of an increase in price and/or advertising time for one channel will instead turn to another of the operator's channels. This operator will set higher prices and/or sell more advertising time than would single-channel programmers of the same channels. The extent of the price increase is limited by the extent to which viewers may be lost to program services controlled by other operators. The ability to internalize the effects of audience diversion due to price or ad-time increases is one reason why programmers may want to combine their services to be sold to viewers as a package.

There are other reasons, pertaining to both demand and supply, that may make multichannel operations more profitable than single-channel operations. Owen and Greenhalgh (1986) found scale economies with respect to both the number of channels and the number of subscribers. The savings realized by programming two channels jointly, sharing studios, and joint management may be why AM-FM radio combinations appear to be more profitable than otherwise comparable AM and FM stations run separately (Anderson and Woodbury, 1989).

Bundling

In addition to cost considerations of the type just listed and the possibility of recapturing viewers lost to price or commercial-time increases, the practice of bundling—packaging channels (or other commodities) together and selling them as bundles—may also encourage multichannel operations. Stigler (1963), in his analysis of block booking in the motion picture industry, was the first to recognize that if buyers differ significantly in their relative valuations of products, then profits may be increased if products or services are packaged to be sold together rather than separately. He saw that selling products in bundles and eliminating the option of purchasing the components by themselves might allow the seller to realize some of the benefits of price discrimination (selling individual products to different buyers at different prices based on willingness-to-pay) in a situation in which the need to charge a single price would otherwise render this impossible.

In Table 4.3, the values in the cells represent the dollar amount each viewer is willing to pay to add the respective channels to the menu of viewing options. If a single firm sells A and B individually, A will be priced at $5 and generate revenues of $10 from sales to viewers 1 and 2. Channel B will be priced at $7 and will sell only to buyer 2, because the price would have to be lowered to $3 to sell to both viewers and total revenue would decline to $6 if both bought at this price. Priced individually, channels A and B will generate a maximum of $17. But if channels A and B are sold as a package, both viewers will buy at a price of $11, generating revenue of $22, $5 more than could be produced by selling the component channels on a stand-alone basis.

Mixed Bundling

Adams and Yellen (1976) extended Stigler's bundling model to show that there are circumstances in which profits can be further increased by offering buyers the choice of the bundle or one or more (but not all) of the bundle's components, a *mixed bundling strategy*. Wildman and Owen (1985) applied the Adams and Yellen framework to video services and examined the stability of bundling in a competitive environment.

Table 4.4 depicts a situation in which the most profitable pricing strategy is to offer viewers the options of A and B as a bundle or A by itself. As in Table 4.3, the profit-maximizing prices for channels A and B sold individually are $5 and $7, respectively, and revenue is $17. The profit-maximizing price for A and B sold as a bundle is $9, and the corresponding revenue is $18. However, if viewers were offered the option of A and B together for $12 or A

Table 4.3 Willingness-to-pay: simple bundling maximizes profits

	Viewers	
Channel	1	2
A	$8	$5
B	$3	$7

Table 4.4 Willingness-to-pay: mixed bundling maximizes profits

	Viewers	
Channel	1	2
A	$7	$5
B	$2	$7

by itself for $7, viewer 2 would purchase the bundle and viewer 1 would buy only channel A, and revenue would increase to $21. Another pricing strategy producing the same result would be to offer viewers channel A for $7 with the option to buy channel B for $5 if they purchased A. This, of course, is the manner in which basic services and optional program tiers are offered to cable subscribers.

Although mixed bundling is more profitable than simple bundling for the demands represented in Table 4.4, simple bundling is more profitable for the situation depicted in Table 4.3. The difference is that the viewers of Table 4.3 are in fairly close agreement on the value of the two channels as a bundle but differ substantially in their valuations of the components; the viewers in Table 4.4 differ significantly in their valuations of the components and the bundle. In the second situation, more is sacrificed in reduced payments from viewer 2 when the price of the bundle is lowered to persuade viewer 1 to purchase it than is gained from the extra payment viewer 1 is willing to make to have both components instead of just channel A.

The welfare implications of bundling are ambiguous. By aggregating the demands of viewers who differ in their willingness-to-pay for different services, bundling may make it possible to supply program services that could not be supported on a stand-alone basis. Suppose, for example, that it costs $11 to produce the programs for channel A and $8 to produce the programs for channel B. If viewer demands are as shown in Table 4.4, neither program would be produced if pricing were on a stand-alone basis only. Both programs would be supplied if they were sold as a bundle, which clearly increases economic efficiency. In some situations, however, bundling or mixed bundling may prevent purchases that otherwise would be made with stand-alone pricing (Wildman and Owen, 1985).

To date, cable has been the dominant provider of multichannel service and has faced competition from single-channel, over-the-air programmers and the VCR, which may be viewed as a single channel programmed by viewers. The competitive equilibrium in this situation has been fairly stable, although cable has been making inroads into the broadcast audience. A growing MMDS industry and the possibility of multichannel DBS service suggest that there will be competition among multichannel services in many areas in the future.[15] Whether more than a single multichannel service can survive in a competitive market in the long run will depend on economies of scale and scope in the provision of multichannel service, how these economies differ among delivery technologies, and the heterogeneity of viewer preferences. Given the high fixed costs associated with the content of media products generally and of multichannel distribution technologies in particular, an equilibrium with price competition among identical multichannel services may not be stable. More likely is the development of multichannel services with differentiated mixes of programs, each service appealing to a distinct viewer group whose programming tastes would only weakly correlate with the preferences of other viewer groups.

New Directions in Program Choice

More recent work on program choice includes models developed by Noam (1987), Waterman (1990), and Wildman and Lee (1989), as well as an extension and empirical application of Beebe's framework by Spitzer (1991). We also present here a new program choice analysis of program stripping. The new work departs from the work in the Steiner and Spence-Owen traditions in that it addresses issues other than structure-performance concerns. Noam and Spitzer study more direct forms of government intervention in program selection than were contemplated in earlier studies. Wildman and Lee consider program repetition as an intertemporal dimension of program choice. Waterman, and Wildman and Lee, look at program quality as a dimension of program choice.

Intertemporal Issues

To this point we have focused on the scheduling of programs within a single viewing period, and have ignored ways in which programming strategies differ in how program selections vary over time. Yet when different channels' program choices over time are examined, important differences that cannot be captured in the single-period representation are found.

Wildman and Lee (1989) calculated the average frequency with which individual program episodes were shown (the first showing plus all repeat showings) on a daily and a monthly basis in the January 1989 schedules of the Chicago affiliates of the 3 major broadcast networks, 4 superstations, 19 basic cable services, 8 pay cable services, and a pay-per-view network (see Table 4.5). The averages mask variation within categories, especially within basic cable, where the figures for broad appeal services such as Turner Broadcasting System (TBS) and the USA Network are similar to those for the superstations and network affiliates. Yet the range of variation, even in the averages, is striking. Pay cable networks such as HBO will show a program unit, such as a movie, almost four times a month on average; broadcast stations typically show their program units only once.[16]

Stripping. Stripping, the practice of scheduling the same program series at the same time on consecutive days, is another type of repetition, although here the unit repeated is a series in a particular daily time slot, not specific program episodes. Outside of prime time and weekends, most of the programs broadcast by television stations are episodes of stripped series—either off-network series previously shown on the networks during prime time or first-run syndicated game shows, soaps, talk/interview shows, and news programs. The same types of programs are also frequently stripped on basic cable channels.

The distinction between the cost of a program and programming

Table 4.5 Average showings per episode

Service	Monthly	Daily
Network affiliates	1.0	1.0
Superstations	1.0	1.0
Basic cable	1.7	1.3
Pay cable	3.9	1.5
Pay-per-view television	19.3	3.2

Source: Wildman and Lee, 1989, pp. 16–17.

costs is essential to understanding the economics of stripping. Programs are obviously a major expense and usually the primary focus of financial analyses. However, substantial resources are invested in selecting programs and developing schedules. Television stations and networks hire highly paid and highly trained staffs to conduct audience research and plot programming strategies. Specialized and costly consultants are also retained to assist with these tasks. As with any other category of expenditure, there are limits to the resources that profitably can be committed to programming. Profit maximization requires that the budget for programming be increased as long as the extra outlay on scheduling strategy generates at least its equivalent in expected increases in revenue.[17] Other things being equal, the resources committed to programming should reflect the size of the audience at stake. More care and expense should be committed to the selection and scheduling of programs broadcast in a market with a million viewers than in a market with only 10,000 viewers. In the larger market a dollar spent on programming can influence the viewing choices of 100 times more viewers than in the smaller market. Even though diminishing returns to programming expenditures will exist in both circumstances, many more dollars will be committed before the expected return on an extra dollar spent on programming falls to a dollar in the market with a million viewers than in the market with 10,000 viewers. Therefore, television stations in large markets have larger programming staffs than stations in small markets, and networks spend more on programming than do stations.

Just as the geographic distribution of viewers and the reach of a channel should influence the resources committed to programming, so should the temporal distribution of the audience affect how programming resources are allocated to the development of schedules for different time periods. Most television viewing occurs during prime time. Prime time also accounts for the bulk of television advertising expenditures. We should expect networks and stations to concentrate their programming efforts on prime time and spend less, on a per period basis, on the selection and scheduling of programs for other dayparts.[18]

The difference in programming resources allocated to prime time and to other dayparts is reflected in the unit programmed. Prime-time hours are programmed daily, and a different schedule is developed for each day. Programming resources are conserved in other dayparts by making program choices apply to all weekday schedules. This is done by selecting previously recorded series that can be stripped or by scheduling first-run syndicated programs such as game shows, talk shows, and soap operas. The per episode cost of each of these first-run genres is low compared with that of prime-time programs and, owing to the nature of the program, new episodes can be produced on a daily basis versus the week to ten days required to film an episode for a prime-time series.

Program repetition. Repeating programs within a schedule can reduce programming costs by limiting the number of program selections. The incentive to repeat programs in a channel's schedule is also influenced by audience share. The smaller the share of audience that a channel can attract with the first showing of a program, the greater the financial appeal of repetition as a programming strategy (Wildman and Lee, 1989). The average number of competitors per audience member is greatest for over-the-air television and smallest for pay-per-view television. The average repetition figures in Table 4.5 are consistent with this hypothesis. In general, share of audience will be inversely related to the number of channels competing for the same audience. Television stations compete with other television stations for the 45 percent of the audience that does not subscribe to cable and with cable services for the rest of the audience. Basic cable services compete with broadcast stations for all of the cable audience and with pay channels for the 80 percent of cable subscribers that take pay services. Pay services compete with all basic channels and all broadcast stations in any given market. Pay-per-view channels also compete with all basic channels and with over-the-air services. We do not have data on subscriber overlap between pay-per-view and pay channels.

A simple two-period example illustrates why the financial attractiveness of program repetition increases as a channel's share of audience declines. Assume that n television channels serve a market and that initially each channel produces a different program for each of the two time periods. The programmer of one of these channels, channel I, is weighing the benefits and costs of showing I's first-period program again in the second period. The benefit is the savings in expenditures if one program is produced (or bought) instead of two. If each program costs x dollars, then the benefit of showing the same program twice instead of two programs once is x dollars. The cost of showing the same program in both periods is the value to channel I of viewers who watched the program during the first period and will watch something else if they are still viewing in the second period. In the second period the repeated program can compete for only that portion of the audience that did not see it when it was first shown. Therefore, the audience for the repeated program is likely to be smaller than the audience for a substitute program that has not been shown before. This reduction in potential audience is the opportunity cost of repeating the program.

Suppose that all channels are ad supported, that the audience is the same size in each period, and that channel I expects to get one nth ($1/n$) of that portion of a period's audience that has not seen the program before. Advertisers pay \$100 for each period's audience. Assume further that all viewers watch television during both periods. (The continuous viewing assumption merely simplifies the example. The qualitative nature of the results would be the same if viewers watched only part of the time, as long as there was some audience carry-over from the first period to the second.) In deciding whether to repeat a single program or to show a different program each period, channel I's programmer compares the expected profits with each strategy. The expected profits with no repetition, $\Pi(\mathrm{NR})$, are

$$\Pi(\mathrm{NR}) = \$200/n - \$2x,$$

which is two times the expected profit of programming each period individually.

The expected profit of repeating the first period's program in the second period, $\Pi(\mathrm{R})$, is

$$\Pi(\mathrm{R}) = \$100/n + \$100(n - 1)/n^2 - \$x,$$

where $\$100(n - 1)/n^2$ is the revenue expected for the program when it is repeated in the second period.

Expected second-period revenue is $\$100(n - 1)/n^2$, because $1/n$ of all viewers are expected to see the program in the first period, leaving $(n - 1)/n$ of the viewing population as the potential second-period audience, and the channel expects to get one $1/n$th of this potential second-period audience.

Subtracting $\Pi(R)$ from $\Pi(NR)$, we see that $\Pi(NR)$ is greater than or less than $\Pi(R)$ as $\$100/n^2$ is greater than or less than $\$x$. $\$100/n^2$, the amount by which the revenue from showing a different program in each period exceeds the revenue from showing the same program twice, falls rapidly as n increases and approaches zero if n is very large. Clearly, there is a value of n at which it becomes profitable for at least one channel to switch from a strategy of scheduling different programs in each period to repeating a single program in both periods. For example, suppose that x is 20. Then if the market had two channels, both would schedule a new program each period because $\Pi(NR)$ is \$60, which is larger than $\Pi(R)$ of \$55. But at least one channel would find it profitable to show the same program in both periods if there were three channels, because $\Pi(NR)$ would be \$26.67, which is less than $\Pi(R)$ of \$35.55.

When the first channel switches to repeated programs, this will reduce the competitive pressures on the remaining channels because they will each be one of just $n - 1$ competitors for the viewers who watched the repeated program in the first period. As the number of channels increases, however, more and more channels will find it profitable to switch to repeated schedules (Wildman and Lee, 1989). This logic applies to pay services as well as to ad-supported services and to competitive as well as monopolized industries.[19]

Government Influence

The performance of the television industry in satisfying viewer preferences has been the primary public policy concern addressed by program choice models. Attention has focused on the influence of structural variables, such as the number of channels or the degree of concentration in the control of broadcast channels, on the efficiency with which television satisfies viewer preferences. To the extent that government intervenes to affect program options in these analyses, it does so by altering these structural characteristics. Yet the government also influences television programming more directly. It requires commercial broadcasters to provide noncommercial programs that serve a vaguely defined public interest.

It regulates obscenity and nudity on the air and it restricts liquor and cigarette advertising. And it allocates many broadcast licenses for the use of the nonprofit stations in the Public Broadcasting Service. (Public channels in the United States are programmed by independent agencies; in many other countries public channels, which may be the only channels, are programmed directly by the government.)

Noam (1987) formally incorporates direct government influence in a program choice model. The model is in the public choice analytical tradition, considering the political trade-offs as well as the economic consequences of regulations and legislation affecting commerce. Noam's wide-ranging analysis covers privately operated monopoly and competitive broadcast industries, government-operated broadcast systems, and mixed public-private systems such as those in the United States, Canada, the United Kingdom, and other European countries.[20]

We focus here on Noam's analysis of mixed systems.[21] In particular, we are interested in his analysis of the interdependence of the programming strategies of public and private broadcasters. His analysis of a mixed broadcast system with private and public channels generates three primary predictions. First, the program biases of commercial broadcasters create political demands for publicly sponsored programming targeted to minority audiences not served by commercial broadcasters. Second, private broadcasters may respond to public broadcaster programs targeted to minority audiences by becoming even more majoritarian. Third, a proliferation of private broadcasters may undermine the political support for public broadcasting.

Noam, like Waterman (1990), uses a one-dimensional spatial representation of program variety; a line represents the range of potential programs and the range of viewer preferences among program varieties. Thus, in a sense, program choice analyses have returned to the spatial roots of the original Steiner model, albeit in a much more sophisticated form. For our purposes, a Beebe-type characterization of viewer preferences is sufficient to illustrate Noam's results (Beebe, 1972, 1977).

Assume that there are three viewer groups: 1,100 type 1 viewers, who prefer type 1 programs and will watch nothing else; 400 type 2 viewers, who prefer type 2 programs, will watch type 1 programs if necessary, but refuse to watch type 3 programs; and 200 type 3 viewers, who prefer type 3 programs, will watch type 2 programs if necessary, but will not watch type 1 programs. Broadcasters earn advertising revenues proportional to their audiences. All programs cost the same. When two or more channels show the same type of program, the audience for that program type is divided evenly among them. Given these assumptions, profit-maximizing strategies are strategies that maximize audiences.

Viewer group	1	2	3
Group size	1,100	400	200
Viewer program preferences			
First choice	1	2	3
Second choice	nonviewing	1	2
Third choice	nonviewing	nonviewing	nonviewing

Initially, there are only two television channels (owing to a limited number of television licenses), even though the aggregate audience is large enough to support more. Each channel is operated by a separate commercial broadcaster. The profit-maximizing strategy for two competing commercial broadcasters is obvious. Each will schedule type 1 programs and realize an audience of 750 viewers, half of the total of type 1 and type 2 viewers. The 1,100 type 1 viewers will receive their first-choice programs, the 400 type 2 viewers will get their second choice in programs, and the type 3 viewers will be provided no programming they deem acceptable. If type 3 viewers feel very strongly about the failure of commercial broadcasters to provide them with the programs they want, politicians or other government officials may undertake to supply type 3 programs with publicly controlled resources to curry the political support of type 3 viewers. Thus type 3 programs may be provided on a publicly supported channel.

Assume that the government does respond to the demands of type 3 viewers and creates a publicly supported channel with type 3 programs. This will affect the programming strategies of any new commercial channels. For example, given a public channel showing type 3 programs, a third commercial channel would provide type 1 programs and share the type 1 and type 2 viewer groups with the first two commercial channels. An audience of 400 type 2 viewers would be smaller than one-third of the combined group 1 and group

2 audience of 1,500. In the absence of the public broadcaster with type 3 programs, however, the third commercial channel would have provided type 2 programs to pick up an audience of 600 type 2 and type 3 viewers. This illustrates Noam's second prediction: when public channels' programs are oriented toward minority-taste audiences, commercial broadcasters will respond by providing more programs for majority tastes. Note that viewers' willingness to watch second-choice programs is critical to this prediction.[22] If viewers refused to watch programs other than their first choices, as in Steiner's original model, public broadcasters' influence on commercial programming would be much less.

Continued expansion in the number of commercial channels illustrates Noam's third prediction. A fourth commercial broadcaster would find it most profitable to provide type 2 programs, even with a public channel that provides type 3 programs. Because type 3 viewers do get some enjoyment from type 2 programs, the type 2 programs of the fourth commercial broadcaster would weaken the political support for a type 3 publicly supported station somewhat. To what degree would depend on how well type 3 viewers think type 2 programs substitute for type 3 programs. If the number of commercial channels continues to increase, eventually at least one commercial broadcaster will find it most profitable to provide type 3 programs and share the type 3 audience. In this example, 15 commercial channels are sufficient to guarantee that one will provide type 3 programs. At this point, the political rationale for the public channel disappears.

The parallel between Noam's third prediction and the plight of public broadcasting in the United States today is quite compelling. Cable networks proliferated during the 1980s, and many began offering programs on topics once considered the exclusive realm of public television. In fact, cable networks purchased programs that previously would have been seen on member stations of PBS.[23] Noam's public choice analysis suggests that it is no coincidence that public television faced severe threats to its federal funding during this period.

Recent work by Spitzer (1991) is consistent with Noam's interpretation of the role of the government in influencing television programming. Spitzer used data collected by the Congressional Research Service to examine government policies designed to promote ownership of broadcast stations by members of racial minority groups. Such policies include favored tax treatment on stations sold to minorities and preferential status in the process for awarding new broadcast licenses. According to Spitzer, government policies may be effective in compensating for the lack of programming preferred by racial minorities because minority owners are more likely to target their programs to minority audiences than are nonminority owners. Limitations of the data set preclude a high degree of statistical confidence in this conclusion, as Spitzer points out.

Program Choice and Program Quality

A common assumption of program choice analyses is that the costs of programs to station and network programmers are fixed. In most of these analyses all programs cost the same. (The Spence-Owen discussion of the bias in favor of programs with small budgets is an exception.) This assumption was employed for analytical convenience and because program diversity, rather than cost, was the primary focus in most cases. In reality, the people who select programs for stations and networks also make decisions that affect the costs of the programs they air, either directly, by setting budgets for programs they produce for themselves, or indirectly, by deciding how much they are willing to pay for programs. In the second case, programmers' willingness-to-pay for programs is reflected in the budgets of the programs supplied to them by producers in Hollywood and elsewhere.

The program budget side of program choice is worth exploring if the enjoyment viewers receive from watching television is affected by the budgets of the programs they watch. One would expect that viewers will find programs with large budgets more appealing than programs with small budgets, because program producers will spend the additional production dollars on things that viewers like. Spending more to get more popular actors, better script writers and directors, and more sophisticated special effects will generally result in a more popular program or film.[24] If it were not possible to increase a program's audience appeal by increasing its budget, profit-maximizing producers would never produce the large budget productions ($1,000,000 per hour and more) we see on prime-time network television. If, say, a $700,000 program would be just as popular, networks would demand the latter.

In the Wildman and Lee (1989) model of intertemporal program choice, equilibrium program budgets vary in response to changes in market structure.[25] Wildman and Lee showed that if increasing the number of competing channels makes it more difficult for any given channel to increase its audience (or what its audience is willing to pay) by showing more expensive programs, then stations and networks will respond to an increase in the number of competitors by producing or buying less expensive programs. Increased competition makes it more difficult to attract new viewers by spending more on programs if, as in the Spence-Owen model, each channel's programs are differentiated somewhat from those of its competitors, so that some viewers always find a new channel's programs better suited to their specific tastes than the programs of the channels that were available before. Viewers who are happier with the programs they are watching will be less inclined to join the audience of a channel that is trying to attract new viewers by showing more expensive programs. As a result, the (marginal) return on a dollar invested in programs will fall, as will the profit-maximizing budget.

In Figure 4.17, MR_1 is a marginal revenue schedule that gives the additional (expected) revenue generated (from increased advertising sales or higher viewer payments) in period 1 by a dollar increase in a single program's production budget as a function of the size of the budget. MR_1 falls from left to right, reflecting diminishing returns to investments in viewer appeal. From a viewer's perspective, the difference between a $100,000 program and a $200,000 program is much more apparent than the difference between a $1,000,000 program and a $1,100,000 program. The return to a dollar added to a program's budget falls to a dollar when the budget has been increased to B_1^*, which is the profit-maximizing budget. Revenue may be increased further by increasing the budget beyond B_1^*, but once the cost of generating the additional revenue is subtracted out, profits on the program will have fallen.

A new technology, or change in the regulatory regime, makes possible an expansion in the number of competing channels in period 2. The new channels are differentiated somewhat from those available in period 1. Programs are offered that some viewers like

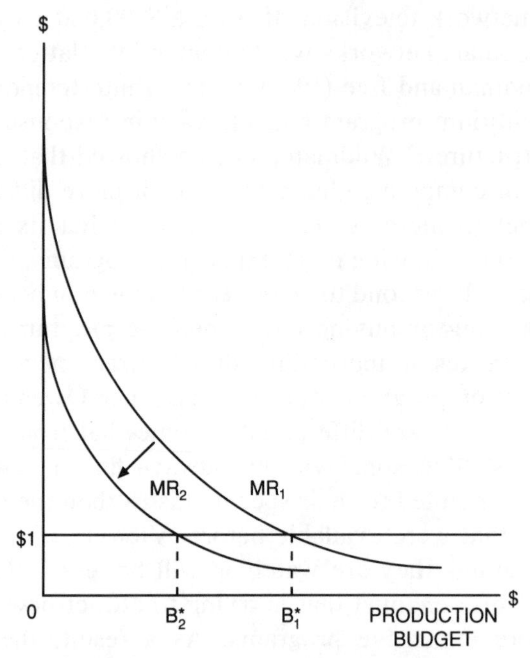

Figure 4.17 The effect of entry on production budgets

more than the programs on the channels available in period 1. Because viewers' preferences in program types are better served, the number of new viewers a channel can attract from its now more numerous competitors with any given increase in its budget for programs will fall. This is reflected by an inward shift in the marginal revenue schedule from period 1's MR_1 to MR_2 in period 2. The equilibrium budget falls to B_2^*.

Budget cuts by the three major broadcast networks beginning in the mid- to late 1980s exemplify this type of response to new competition. Although some cuts reflect the fat in previously bloated staffs, others—such as the move toward less expensive (and less popular) news-format prime-time programs and the elimination of many regional news bureaus—have affected the quality and viewer appeal of network programs.

In a competitive television market there is a trade-off between program diversity (which increases with the number of competitors) and production quality (which increases as program budgets increase). This is quality measured in terms of viewer appeal and not by any specific aesthetic criteria. This trade-off complicates the application of program choice theory to television policy. Diversity, the "good" produced by encouraging competition, is now seen to have a higher cost than just the possibility of wasteful investments in programs that are too similar to each other to generate viewer benefits commensurate with their production costs. With the traditional models of program choice, diversity generally increases as the number of competing channels increases, and this benefits viewers. The variability of quality in the more modern models raises the possibility that viewer welfare may decline as diversity increases, because of the associated decline in production quality.

Wildman and Lee's model does not permit welfare comparisons of competitive equilibria characterized by different combinations of program diversity and production quality. But Waterman (1990) demonstrates a tendency for a competitive pay-television industry to oversupply diversity and undersupply quality. He finds that an advertiser-supported television industry also may oversupply diversity, but whether quality is over- or undersupplied depends on the amount advertisers pay per viewer for the audiences produced by programmers. Waterman employs a spatial representation of the demand for product quality, with viewers' preferences and variation in program content represented by locations on a line, as does Noam's (1987) model and the original Hotelling (1929) spatial analysis. As Waterman points out, his results may be dependent on this particular demand specification. The generality of this result may be determined as more quality-variable program choice models are developed.

A competitive television market is more likely to supply more than the welfare-maximizing number of channels in a quality-variable world than in the quality-constant worlds of the Steiner and Spence-Owen models. The competitive equilibrium number of channels may exceed the social optimum in the Steiner and Spence-Owen analyses because there are circumstances in which a new channel contributes more to its operator's profits than the value of its contribution to viewer satisfaction. This is most likely when competing channels are close substitutes. New channels will be introduced as long as the audiences they attract generate revenue sufficient to cover program costs and operating expenses. However, when new channels are close substitutes for preexisting channels, viewers derive little additional value from them. The new channels simply "cannibalize" older channels' audiences and revenues. This is most obvious in the Steiner models, where new channels may carve up the audience for a given program type while contributing nothing to viewer satisfaction. Viewers who stay with their original channels are no worse off, however, because they still get as much enjoyment from watching them as before. By contrast, when quality is a variable, established channels respond to competition from new channels by using less expensive programs, which makes the viewers they retain worse off.

Summary

The biases of advertiser support revealed by the Steiner and Beebe analyses of Chapter 3 are also evident in the more sophisticated models reviewed in this chapter. In addition, we see that pay television is subject to the same biases, although to a smaller degree. Relative to the viewer benefits provided, both advertiser-supported television and pay television have three biases: against programs that cater to minority-interest tastes, against expensive programs, and in favor of programs that produce large audiences. These biases are less pronounced for pay television, because the intensity of viewers' preferences is reflected in the prices they pay. From a social welfare perspective, there is no unambiguous ranking of alternative ways to organize a television industry. Spence and Owen (1975, 1977) compared the welfare benefits produced by a pay monopoly, a competitive pay industry, an advertiser-supported television monopoly, and a competitive television industry supported solely by advertising. They showed that a competitive pay-television industry always does better than a pay monopoly. Which of the three options other than pay monopoly does best depends on the strength of viewers' preferences for diverse programming. An ad-supported monopoly does best if tastes are relatively homogeneous, because service is priced at marginal cost, which is zero, and a monopolist will not provide an excessive number of highly similar programs. As the preference for diversity increases, the welfare-preferred industry structure shifts first to advertiser-supported competition and then to competition with pay support.

The political system responds to the biases of commercial television by sponsoring public television channels with programs that cater to the preferences of viewers not well served by commercial channels. But as the number of channels employed by commercial television increases, new commercial services develop that provide programs increasingly similar to what is offered on public channels. This weakens public television's political support, which is derived from its differentiated programs.

The programming strategies of commercial channels change as the number of channels increases. As the audience per channel falls, the size of program investments that can be supported by any given channel must fall as well. Commercial channels will also look for ways to reduce programming costs, perhaps through stripping and through increased repetition of programs within their schedules. Repetition has the advantage of reducing program costs. Its disadvantage is that the potential audience for the second showing of a program is reduced by the number of viewers who saw it when it was first aired. Thus programs draw smaller audiences the second time they are shown than would new programs shown in the same time slots. Because audience shares diminish with increasing competition, the lost-audience penalty of program repetition falls, so that the incentive to repeat programs increases.

From the viewer's perspective, advertising is one of the costs of programming. Many cable channels are supported by a combination of viewer payments and advertising. This may be seen as a combination of direct and indirect viewer payments in terms of the amount of advertising time viewers must endure to watch a program. Pay cable channels are supported entirely by viewer payments. The ability to tap viewers' willingness-to-pay is an advantage to cable and other delivery media that can charge viewers; over-the-air broadcasters are supported entirely by advertiser payments. The existence of pay-only channels and channels supported by a combination of viewer and advertiser payments on cable may be seen as another form of price discrimination. Viewers with the lowest tolerance for commercials pay more for commercial-free channels.

Multichannel television services, such as cable and MMDS, bundle channels together in packages that are sold at a single price. Bundling may increase revenues and profits when viewers differ in how much they value the various channels in a bundle. Viewers who value some channels in a bundle highly and value other channels considerably less are balanced against viewers who value the second set of channels more than the first. The price of the bundle reflects the sum of viewers' willingness-to-pay for the channels they value highly and what they are willing to pay for the channels they value less. By getting viewers to pay at least a small amount for channels they value relatively little, broadcasters extract some payment from viewers who would not be willing to pay the price that would maximize profits for the same channels sold separately.

5 / Network Economics

Unlike many other kinds of networks, such as airlines, railroads, highways, telephone systems, and electric power grids, which carry people, goods, or services from one point to another, mass media networks carry messages from one point to many. The oldest electrified media networks are the wire services that distribute news bulletins to subscribing newspapers. The first true networks for mass media connected radio stations and were formed by David Sarnoff and William Paley during the Great Depression. Broadcast television networks followed after World War II, and the first cable television networks were formed in the mid-1970s.

Transportation, electric power, and telephone networks face the same economic problem: how to minimize the cost of connecting many points of stochastic, interdependent demand for service, given the cost characteristics of line hauls and switching. Mass media networks face a simplified version of this classic problem, but their primary challenge is to exploit the economies of scale arising from the high "first-copy costs" of media messages that are of general interest. Messages of general interest are "public goods." The greater the audience or potential audience for these messages, the lower the cost per consumer. The production of mass media messages involves a trade-off between the savings from shared consumption of a common commodity and the loss of consumer satisfaction that occurs when messages are not tailored to individual or local tastes.

Advertising is central to broadcast networks because the economic forces favoring mass consumption of media messages are reinforced by the simultaneous production of audiences for sale to advertisers as a by-product. Advertisers of products for which virtually all consumers are potential purchasers want to reach very large segments of the population. Advertisers of products with a narrower group of potential purchasers seek more specialized audiences. Because of the economies that exist in producing large audiences, and because there is less aggregate demand for undifferentiated audiences, advertisers generally pay less per member of the audience for large audiences than for small. Large audiences also tend to be less geographically or demographically specialized, and thus they often result in wasted exposure for advertisers. But very large audiences avoid the problem of duplicate exposure that arises when an advertiser purchases access to a series of smaller audiences.

Competition for advertising often serves viewers' interests, because the media compete for audiences by increasing the attractiveness of their programs. But this usually increases program expense, as well. Both competition for scarce inputs and the utilization of more inputs make attractive media products more expensive.

The existence and structure of mass media of all sorts (including magazines and newspapers) can be understood in terms of three fundamental forces:

- the cost savings (but lessening of appeal to individual or local taste) that result from mass consumption of common messages;
- the demand by advertisers for access to audiences with a potential demand for their products;
- the competition among the media in terms of expenditures on attractive programming to acquire subscribers and to generate audiences for sale to advertisers.

These forces produce diverse media with different specialization and reach and different mixtures of support from advertisers and consumers. The television broadcast networks (and before them, the radio networks) were required at first by technology and later by law to be supported entirely by advertising. In addition, government restraints on entry severely limited both the number of competitors and, as a result, the diversity of programming. As it happened, consumer taste and advertiser demand favored the development of national rather than local messages and audiences, at least with respect to entertainment programming. (This is in contrast to the daily newspaper industry, where a somewhat different balance of these forces produced not a national press but a series of local monopolies.)[1]

This chapter describes the economic and regulatory forces that have influenced the past and will shape the future of television networks in America. From a powerful duopoly of NBC and CBS, the networks became an equally powerful oligopoly with the addition of ABC as an effective competitor in the early 1970s. Regulatory reform and evolving technology, including the advent of cable networks, then ushered in a period of intense competition and gradual decline.

Network Basics

The broadcast networks act as brokers and consolidators for local affiliated television stations in the business of selling access to audiences. The price of a 15-second or 30-second commercial varies with the size of the audience, and is usually quoted in the trade as the price per thousand viewers (or households). The process by which the networks acquire the ability to sell audiences to advertisers is as follows. Television programs are purchased by the networks from the program production industry, or they are produced by the networks themselves. These programs are delivered to individual television stations across the country in return for access to local audiences. The network thus serves as an intermediary between local stations and advertisers and program producers, and (to the extent that it performs the function of program "selection") between local stations and viewers. Cable networks perform a similar function. Networks must deal with three groups, advertisers, viewers, and program producers, and must also work with their local affiliates.

The Advertising Market

Despite the prominence of television commercials as a form of advertising in the United States, only about 8 percent of total advertising expenditure, or about 14 percent of all national advertising expenditure, goes to broadcast network television. Since the early 1980s these percentages have been decreasing, as is shown in Figure 5.1.

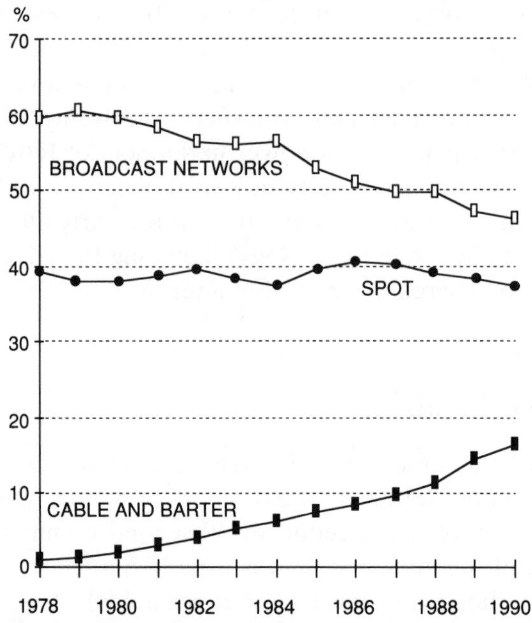

Figure 5.1 National television advertising, 1978–1990. Data from *Kagan Media Index,* 1990, 1991.

There are a number of more or less good substitutes for network advertising: spot television advertising, advertising on basic cable networks and superstations, network and spot radio, national magazines, direct mail, billboards, and newspapers. None of these substitutes reaches as large an audience as network television, but many advertisers prefer a smaller audience to a larger one if that audience contains a greater number of likely customers. (Firms whose sales outlets are geographically concentrated provide the most obvious example.) By careful program selection, the networks do try to supply audiences that have the characteristics that advertisers want. Because of their buying power, women between the ages of 18 and 49 are a desirable audience. Network television's principal differentiating characteristic is that its large audiences do not contain the duplication that would be present in an accumulation of smaller audiences.

Advertisers can buy access to network audiences in a number of ways. They may sponsor an entire program or series of programs—a practice that was standard in the 1950s but is now quite rare, except for special event programs in the high culture or public affairs category. More common is the purchase of "participations," or partial sponsorship of programs selected by the network. A large network television advertising account will typically buy, late in the spring, a package of commercial minutes spread over a number of programs throughout the next television season. As much as 70 to 80 percent of network time is committed in that "up-front" market. Time not sold in the up-front market is sold during the broadcast season in the "scatter" market. Some commercial minutes may remain unsold until days or even hours before they are broadcast. Commercial minutes that cannot be sold are filled with public service announcements or network promotional announcements. Network promotions are not merely filler, however; they are crucial to the success of the networks' upcoming programs. The networks' ability to promote their own programs to large audiences is an important ingredient in their continued relative success vis-à-vis their smaller rivals. Sales resulting from heavy promotion similarly explain the success of downstream sales of national motion pictures.

Advertisers seek to reach potential purchasers of their products. Because these purchasers are not distributed uniformly across the population, each advertiser will have different preferences for audiences. The most prominent of these preferences are geography and demographic composition (age, sex, income, and education). Many advertisers are concerned with the timing of their advertising campaigns, because they need to coordinate exposure on different media and to coordinate advertising with retail promotions and with production and inventory activities; few if any national television advertisers are concerned with which day of the week their commercials appear.

The number of competing stations has no effect on the price of advertising time sold by television stations in a particular area (Fournier and Martin, 1983; Fournier, 1986). Fournier and Martin (1983) conclude that, because additional stations provide alternative sources of supply to advertisers, the absence of price effects may signify that stations are competing in broader markets that include perhaps wider geographic areas and other types of media. The prices of advertising sold by stations and measures of concentration of station time sales do not have the positive relationship that would be predicted if local spot advertising time sold by stations constituted a market within which tacit collusion were possible. Although these results refer to local stations, the implication that advertisers seeking local audiences regard other local media (mainly print and radio) as substitutes for television advertising is presumably applicable to advertisers seeking national audiences as well.

The most immediate substitute for network television advertising may well be other national television advertising. There are three major species of national non-network television advertising: advertising on cable channels, barter-syndication advertising, and advertising in the national spot market.

Advertising on national cable networks such as ESPN has grown rapidly from about $50 million in 1980 to about $2 billion annually in the early 1990s. The 1980s was a decade of phenomenal growth for barter-syndication advertising as well. Barter syndication involves the sale of national advertising by program producers or their representatives. The commercials are inserted into programs that are supplied to local stations, often by satellite. In barter-syndication transactions, a program producer in Hollywood packages a show or a series together with national advertising messages. The producer then sells the show to local stations for a low price (sometimes for free). The local station sells additional commercial time in connection with the show. By 1990 barter-syndication sales reached about $1 billion per year. According to the five-year forecast of Veronis, Suhler & Associates, barter-syndication program buying is expected to reach $2.3 billion by 1993 at a compounded annual growth rate of 20.2 percent.[2]

National spot sales involve transactions between national advertisers (or their agencies) and local stations (or their sales representatives). Such advertising sales bypass the networks. Their advantage to advertisers is that they permit the ad campaign to be tailored to the audience most likely to purchase the product being advertised. Although the cost per thousand viewers is theoretically lower (because of transaction costs) for network than for equivalent national spot campaigns, the network sale generally involves some "wasted" expense: the cost of exposure to audiences in portions of the country where the advertiser has few outlets or otherwise poor prospects (Poltrack, 1983, pp. 294–295). From this perspective, network sales and national spot sales are best regarded as differentiated, substitute products.

There is broad agreement that national spot and network ads are generally good substitutes. Using published rate cards, Peterman (1979) compared the cost of buying exposure to national audiences via network advertising with the cost of purchasing advertising time from stations in the national spot market; he found only slight differences in the costs of reaching equivalent audiences. The FCC came to similar conclusions (FCC, Network Inquiry Special Staff, 1980b). Spot and network prices are interdependent. Increases in the price of network advertising affect advertisers' allocations of budgets between network and spot. The FCC study found that spot television, radio, magazine, newspaper, and outdoor advertising constrain the prices of network advertising and that the prices networks charge for viewer exposures reflect competitive forces.

The National Association of Broadcasters, through its industry code, once set an upper limit on the number of commercial minutes per hour (six minutes per hour in prime time) for network commercial messages, and the FCC implicitly endorsed this decision. The practice, however, was declared to violate the antitrust laws in 1981, and no such agreement exists today. (Following instructions from Congress, the FCC in 1991 did impose limitations on commercials in children's programs.)[3] On the basis of the analysis described below, the NAB Code limits do not appear to us to have been an effective constraint on overall competitive activity (see also Hull, 1990). The expansion in minutes of commercials parallels the increase in hours of programming (see Tables 5.1 and 5.2). Both have occurred primarily during the late fringe (11:00 P.M. to 1:00 A.M.) and weekend dayparts. Table 5.3 gives the average number of minutes of commercials per hour of programming. Unfortunately, the available data do not permit one to compute minutes per hour for prime time for the period just before the NAB Code was invalidated. Minutes of commercials per hour in prime time expanded after 1982.

The length of commercials on network television has become shorter, however (see Table 5.4). There are far more 15-second commercials today than in 1981, partly because three-quarters of the audience can switch from boring commercials by using their remote controls. According to network lore, 15-second commer-

Table 5.1 Minutes of commercials sold by broadcast networks, 1980–1988

Year	Prime time	Total
1980	43,093[a]	123,966
1981	42,153[a]	122,297
1982	23,380	126,987
1983	23,980	134,126
1984	24,192	135,477
1985	24,328	133,240
1986	24,673	132,765
1987	24,805	132,173
1988	25,112	137,146
1990	28,183	142,394

Source: Compiled from data contained in Leading National Advertisers, *BAR/LNA Multi-Media Service*.

a. "Night time" in the source is undefined, but clearly covers more than the current definition of prime time.

Table 5.2 Hours of programming by broadcast networks, 1980–1988

Year	Prime time	Total[a]
1980	3,922[b]	12,859
1981	3,971[b]	13,294
1982	3,432	13,575
1983	3,432	14,249
1984	3,432	14,067
1985	3,432	14,151
1986	3,432	13,894
1987	3,432	14,129
1988	3,432	14,099

Source: Compiled from A. C. Nielsen data.

a. Data are based on Nielsen statistics for the second week of November, February, May, and July in each year.

b. Includes network news time plus prime time.

cials are about 70 to 80 percent as effective as 30-second commercials but cost only half as much, which makes them more cost effective for advertisers.

Program Markets and Syndication

A television station has two external sources of programming material: affiliation with one of the four national networks and the syndication market. The networks distribute programs that have been (1) obtained directly from program producers; (2) supplied by advertisers who buy network time; or (3) produced internally. Stations can acquire non-network fare—first run or rerun—by going directly to syndicators for television series, specials, and feature-length films.

The supply of programming to the networks has been unconcentrated and competitive. Historically, the networks distributed the programs, taking advantage of any economies of distribution. Economies of scale were limited, evidenced by the continued viability of a wide range of firm types and sizes. The rental market for inputs used in program production was well developed, which spread the cost of expensive inputs (such as sound stages) over several producers. Entry by new competitors was easy, which frustrated any attempt by packagers to earn monopoly profits through collusion.

Table 5.3 Minutes of commercials sold by broadcast networks per hour of programming, 1980–1988

Year	Prime time	Total
1980	—	9.6
1981	—	9.2
1982	6.8	9.4
1983	7.0	9.4
1984	7.0	9.6
1985	7.1	9.4
1986	7.2	9.6
1987	7.2	9.4
1988	7.3	9.7

Source: Compiled from A. C. Nielsen data.

Table 5.4 Thirty-second commercials on network television, 1981 and 1987

	November 1981	.November 1987
Commercials aired (number)	3,606	4,667
Commercial minutes	1,802	1,954
Percentage of total		
30-second commercials	92	61
15-second commercials	2	36

Source: Nielsen Media Research as reported in A. Marton, "Ad Makers Zap Back," *Channels,* September 1989, p. 31.

In the 1980s, although the concentration of suppliers increased, perhaps because of the risk-shifting effects of certain FCC and Department of Justice regulations discussed later in this chapter, concentration remained low.

Networks purchase prime-time series, specials, and television rights to sports programs from outside vendors. News, sports, and late night and early morning talk shows are produced in-house. Prime-time series, news, and sports rights account for the bulk of program expenditure. Successful prime-time series are sold as reruns in the off-network syndication market. In addition, Hollywood produces a number of first-run syndicated shows, never shown on a network, that can be purchased by stations. This category is dominated by inexpensive, but increasingly popular, game and talk shows.

The supply of syndicated programming has been at least as unconcentrated as the supply to the networks, and substantial relative growth in first-run syndication has reduced overall concentration in syndication. Any economies from distributing programs, with their public-good characteristics, appear to be quickly exploited. Many firms are viable, each with only a tiny fraction of total syndication production and distribution. Table 5.5 shows levels of

Table 5.5 Share of audience delivered, syndicated programs (percentage of total syndicated program audience)

	1971	1989
First-run programs		
Top 5 syndicators	50	45
Top 10 syndicators	67	64
7 major studios	4	34
Off-network programs		
Top 5 syndicators	55	63
Top 10 syndicators	76	82
7 major studios	33	58
Total syndication		
Top 5 syndicators	43	44
Top 10 syndicators	61	64
7 major studios	23	41

Source: Crandall, 1990f, Appendix A.

concentration in supply of syndicated programming to television stations in 1971 and 1989, based on audience delivered.

The major broadcast networks can easily compete with syndicated programming for two reasons. First, station profits are generally higher with network fare. Talent is available at lower wages for a first-run syndicated show than for a network program, because of union arrangements and because more work can be guaranteed (not just a pilot, or a pilot and thirteen episodes). But talent cost is only one component of the cost of a program. Although the syndicated program has lower talent costs, it may have other costs that are higher. Economies in simultaneous networking confer a competitive cost advantage on network fare because of the effect of sharing program costs over a larger audience and the networks' lower distribution costs. Even those syndicators that take advantage of communication satellites to distribute their shows to stations have higher costs, because the expense of using a satellite is independent of how many stations receive the broadcasts, and fewer stations receive any given syndicated program.

Second, the affiliates in cities with more VHF stations than networks may use network fare because they fear loss of their affiliation agreement. If the threat of cancellation is credible, the affiliate will maintain its affiliation only if it is more profitable than independent status. As an affiliate it can show network and some syndicated fare, but as an independent it can show only syndicated programs. Affiliates remain affiliates because the network supply of programming *as a whole* is more profitable than is syndicated material, though not necessarily on a program-by-program basis. The advantages of affiliation are likely to be far greater for ABC, CBS, and NBC stations, of course, than for stations affiliated with the Fox network, which is much newer. Fox has faced the difficult task of persuading actual and potential affiliates that its network programs will be more profitable than the stations' previous syndicated fare despite circumstances in which that claim appears barely valid.

Even major network affiliates prefer some syndicated programs over programs supplied by the networks. Whenever a network program is available and an affiliate preempts it for a syndicated program, "network coverage" is impaired. "Coverage" refers to the percentage of U.S. television households residing in the markets of network affiliates that "clear" (broadcast) a given episode of a particular program. If each affiliate clears its network's program, each major network can reach over 99 percent of all U.S. television households. (Fox reaches fewer viewers at present.) If affiliates preempt their networks' programs with syndicated alternatives, coverage falls. Between 1980 and 1989, prime-time coverage consistently averaged from slightly over 97 percent to slightly over 98 percent (see Table 5.6). Daytime coverage was generally in the low to mid-nineties. The most popular programs are rarely preempted; the programs with the lowest shares are preempted most often. Preempting is most troublesome for the networks in the late night and early morning dayparts, which are difficult to program (Mandese, 1989).

The extent to which affiliates preempt network programs is revealed by the number of hours preempted as well as by the difference between actual coverage and the theoretical maximum of over 99 percent. According to the *Gallagher Report,* affiliates preempted 18,000 hours of network prime-time programs for one-time-only (OTO) syndicated programs during the 1987–1988 television season. This is consistent with figures quoted by network officials.[4] Averaged over approximately 200 affiliates for each major network, this amounts to approximately 0.6 hours of one-time-only prime-

Table 5.6 Average coverage, entertainment programs, three broadcast networks, 1980–1989 (percentage of television households in markets reached by network broadcasts)

Year[a]	Prime time	Daytime
1980	97.39	94.66
1981	97.38	94.05
1982	97.93	94.40
1983	97.16	93.10
1984	97.20	92.79
1985	98.11	93.79
1986	97.73	93.76
1987	97.97	94.29
1988	97.74	92.96
1989	98.21	93.56

Source: Compiled from A. C. Nielsen data.

a. Coverage is for the last two weeks of January each year.

time preemptions per week per affiliate. The total replacement of a network series with a syndicated series is much less common, especially in prime time, but it does happen occasionally. "Star Trek: The Next Generation" is probably the best-known example of a first-run, syndicated program that is frequently included in the schedules of network affiliates. Star Trek has a network-size budget and is regularly among the top-rated syndicated shows. Paramount had offers for this program from the networks, but chose to syndicate it instead because it thought it could do better in syndication.

One-time-only preemptions have been a contentious subject in meetings between networks and their affiliate groups. The networks claim that these preemptions are becoming increasingly costly, particularly for daytime programs. Although there is great variation in the data, the percentages in Table 5.6 suggest a decline in daytime coverage. The networks have attempted to restructure affiliation contracts in part to provide incentives for affiliates to clear more network programs.

The tendency of affiliates to preempt less popular programs more frequently than highly rated programs reflects the effects of two FCC restrictions on network-affiliate contracts. Networks are not allowed to force clearance of less popular programs by withholding more popular ones (Besen and Soligo, 1973). As it is, the network compensation payments to affiliates tend to be uniform within a daypart. Because affiliates sell advertising time in network programs, the lowest-rated network programs are least profitable to affiliates and are most likely to be preempted. Theoretically, networks might still limit preemptions by requiring affiliates to clear all network programs as a condition of affiliation. Such contractual restrictions would be feasible, because network schedules as a whole are more profitable to affiliates than are complete schedules of syndicated and self-produced programs. However, contractual restrictions of this type are precluded by FCC rules forbidding networks from contracting for exclusive access to prespecified blocks of broadcast time (option time). The FCC thus guarantees affiliates' rights to preempt network programs.

The extent to which viewers and affiliates benefit from preemptions is unclear. Syndicated specials do not always achieve their hoped-for ratings. More significant, an affiliate might prefer a smaller audience with better local demographic characteristics, or a program generating a larger local audience in that market. In these cases viewers and affiliates in other markets are worse off, because the network will invest less money in the quality of those programs that are widely preempted. Affiliates must at least earn more on average from the syndicated specials than they would with the network programs they displace; otherwise this practice would not persist. Excessive preemptions, however, harm the long-term interests of viewers, affiliates, and networks.

Until network programs are broadcast, it is impossible to predict their audience appeal. Programs with comparable budgets frequently differ widely in the numbers of viewers they attract. As a result, even within a daypart, a network's schedule will contain programs that differ greatly in their popularity. The expected audience for a given program will be some weighted average of the audiences for all of the programs in a network's schedule.

The true appeal of a network series is revealed during the course of a broadcast season. This is an advantage to producers of a one-time-only syndicated special. Stations must gauge its potential audience appeal on the basis of its budget and subject matter. Thus stations will evaluate a one-time-only special by comparing its ex ante expected audience to the audiences known to be generated by the least popular network programs. (One-time-only programs such as "Billy Graham Crusades" often have a track record.) It is not surprising that stations should find one-time-only specials to be attractive relative to the lower-rated network programs.

Preemptions of a network's lower-rated programs reduce the average return on its program investments by limiting the audiences for these programs to less than what would be realized otherwise. The profit-maximizing response to reduced return on investment is generally a reduction in the total resources invested. Because it is not possible to predict beforehand which new programs will be most popular, this means smaller budgets for program development. Reduced expenditures on program development will be reflected in less popular network programs overall and network schedules that are less competitive with the programming of cable services and independent stations. Thus, in the long run, preemption may be inimical to the interests of viewers and network affiliates. The behavior of individual network affiliates may be at odds with the interests of affiliates as a group, a state of affairs that could be cured, in principle, through affiliation agreements and compensation arrangements that restricted preemptions. But the FCC prohibits such agreements.

Viewer Behavior

The audience produced by a broadcaster is related to the popularity of its programs. But the total television audience seems to be determined almost completely by exogenous factors, such as seasonal patterns of family entertainment. Consider, for example, average April prime-time television usage, or "HUT" (the percentage of television households viewing any form of television except prerecorded cassettes), between 1953 and 1990. Usage varied between a minimum of 57.3 percent in 1963 and a maximum of 63.3 percent in 1983, with no particular long-term trend.[5] This stability occurred despite drastic changes in the kind and quality of programs offered, the advent of color television, the fact that April is now (as it was not in earlier years) occupied almost entirely by rerun programming, and a great increase in the number of households with televisions and in the number of hours per day during which each household views. Most significant, there has been a great increase in choice. More than 50 percent of all television households are now cable subscribers, and 80 percent own VCRs.

The audience for a program depends not only on the program but on the popularity of adjacent programs (Epstein, 1973, pp. 93–97). With remote control devices, viewers switch channels more frequently, "grazing" among their entertainment options. Despite this channel hopping, a network must pay attention not merely to the popularity of individual programs but to its whole schedule in relation to the schedules of its rivals, because there remains a tendency on the part of viewers to stay tuned to the same channel unless provoked. Thus, because overall viewing is not affected by what is offered, a network can increase its instantaneous audience only at the expense of its rivals, and even then, increases derive more from raising the popularity of an entire schedule than by variations in the quality of individual programs.

A program's quality is measured by its contribution to the audience of the network. Popularity can be increased by a very uncertain process of feeling out public tastes in relation to the offerings of rival networks. There is a role here for the "novel idea," but most quality variation takes the form of providing more of those program "values" or types that appear to be popular on other networks or in feature films. A network can do this by producing more elaborate versions of the same type of program or by bidding away actors and directors from existing programs or films. Either action is likely to increase the cost of the program. Program popularity, quality, and cost are highly correlated.

There are exceptions to this, of course, whenever a popular new program or talent is discovered. At first an unexpectedly popular program will cost less than programs of equal popularity that are older, and whose input factors have been evaluated properly by the market. If the returns to actors or directors from participating in network television are higher than their opportunity costs in alternative activities, scarcity rents for these talented individuals will arise. It is interesting that many popular cable networks have chosen entirely new formats rather than try to compete for viewers on the basis of ordinary network fare. For example, ESPN specializes in sports, CNN in news, MTV in music videos, and the Discovery Channel in nature programs. On the other hand, Fox, the first new broadcast network to enter the business in more than 40 years, has generally chosen conventional format shows, modifying this strategy only slightly to appeal to relatively young audiences.

The allocation of rents among networks, talent, and program suppliers was examined by Woodbury, Besen, and Fournier (1983). They found that suppliers and talent participate in the profits of unexpectedly successful shows. Their empirical findings are "consistent with the view that program suppliers serve as vehicles for transferring rents from the networks to talent inputs" (ibid., p. 351).

Affiliated Stations

Affiliated stations have contracts (in effect, franchises) with the networks, the terms of which are restricted by the FCC in the stations' favor. Originally imposed to protect radio stations from the exercise of network power and to ensure that local stations rather than networks were responsible for program content, these "chain broadcasting rules" are largely ineffectual and often perverse in their effect. The networks try to line up as many "clearances" with affiliated stations as they can. The stations have an incentive to "clear" network programs only if the compensation they receive from the network exceeds the profits they could earn by using other program sources. This opportunity cost is determined by the revenue for spot and local commercials less the cost of originating local programming or acquiring syndicated programming and the selling costs involved, taking into account the effects of this on audience size and on "audience flow" among adjacent programs. Network programs in the past were sometimes carried at a net loss by local affiliates because the programs satisfied FCC requirements (Epstein, 1973, pp. 84–91), but in a deregulated context this phenomenon is much less important (Besen et al., 1984).

Most stations affiliate with a network to take advantage of economies of scale in sharing program costs over large audiences and in selling commercial minutes to national advertisers. These and other benefits of network intermediation outweigh the disadvantages to stations of having to carry a program that is not always the one that maximizes audiences in a particular city. In general, network affiliation is the most profitable choice for a local station to make. Affiliation gives a station the option to carry popular and expensive programs without obligating the station to carry any particular program if something more profitable is available. Most stations consider network affiliation their most important single asset, next to their FCC license.[6] Our own study of preemption activity by network affiliates, discussed earlier in this chapter, suggests that even increased opportunities to buy popular syndicated programs has not resulted in any significant change in network clearances by affiliates of major networks. It may be more telling, however, that network clearances have not *increased*. The number of markets with four or more VHF stations has increased dramatically. Networks have a bargaining advantage when the number of potential affiliates exceeds the number of available networks. Other things being equal, the growth in the number of markets in which networks have a relative advantage should increase the rate at which network programs are cleared. But the emergence of the Fox network has shifted the advantage away from the major networks and toward the stations.

The ability of a network to extract economic rents from local stations is limited by its ability to find other stations in the same market with whom to affiliate. At least 42 cities have more than three VHF stations. It is presumably in these cities that a network's threat to cancel an affiliation contract would be most effective, at least prior to the entry of the Fox network. In West Palm Beach, Florida, intense competition among stations for affiliation in 1988 produced "negative" compensation from a major broadcast network for the first time. Network compensation is the net of two offsetting dollar figures. One of these is the revenue that accrues to the station from the sale by the network of advertising time belonging to the station. The second is the revenue accruing to the network from the station to compensate the network for supplying programming. Compensation historically has been a net payment from the network to the station. But stations also sell their own ads in conjunction with network broadcasts, so there is no reason in principle why compensation could not flow in the opposite direction, as it did in West Palm Beach. Indeed, it is not uncommon for new cable networks to pay cable operators for access in order to generate larger potential audiences and advertising revenues.

In markets with only three VHF stations neither the stations nor the networks, prior to Fox's entry, had any particular bargaining power in this respect, because neither side had a viable alternative. The "strongest" network, of course, has bargaining power by virtue of its attractiveness to stations affiliated with "weaker" networks; similarly, the station affiliated with the "weakest" network will be the affiliate most likely to engage in preemption of network programs. In one- and two-station markets (and, with Fox, in three-station markets), power tends to lie on the side of the stations. The opposite holds in the largest markets, which have four (with Fox, five) or more stations.[7] Because affiliates in markets with more than four or five stations are in larger, and hence more profitable, markets, their profits will be larger than those in three-station markets. This will occur despite their weaker position vis-à-vis the networks. In addition, many of the affiliates in the largest markets are owned and operated (O&O) by the networks. In such cases there is no bargaining problem, although there may be accounting questions.

The networks have been suspected of "hiding" their profits in O&O stations by paying them generous compensation, which serves both to disarm critics and regulators and perhaps to enhance the network's bargaining position with non-O&O stations in smaller, less profitable markets. If a network can hide profits in O&Os, other affiliates may believe the joint profits of the network to be smaller than they actually are. With imperfect information, the other affiliates may settle for smaller compensation than they would if they knew the real magnitude of network profits.

Even in the three-station markets, the FCC rules of affiliation give considerable latitude to local affiliates in dealing with the networks. Let us examine the reasons. The station-network relationship can best be understood in terms of a bargaining problem. Consider the hypothetical case of only one network and a single local station. Each can earn $1 million operating alone (or in the case of the network, operating only in other markets), but together they can earn $3 million in profit. If a bargain of affiliation is struck, each will settle for no less than $1 million. Any division of the remaining $1 million will maintain the contract; for example, $1,999,995 for the network and $1,000,005 for the station or vice versa, or somewhere in between. Both the station and the network are better off by affiliating, and neither has a sensible alternative.

Now consider a second hypothetical case of only one network and two local stations, where collusion by the local stations is effectively prohibited by FCC regulations and antitrust laws. Separately the network and the two stations can each earn profits of $1 million, but an affiliation contract between the network and either station will yield joint profits of $3 million. Clearly, the network will capture the extra $1 million generated by networking. If station A has the affiliation contract and asks for more than $1,000,005, the network will offer station B the contract for $1,000,004. Because this will increase the profit of station B, it will take the contract. This process can be repeated until the network receives $1,999,999.99, the affiliated station gets $1,000,000.01, and the unaffiliated station earns $1,000,000.00. (In the real world, with imperfect information, affiliates will receive more than 1¢ from affiliating with the single network.) If there were two networks and one station, the bargaining positions would be reversed. The station would capture all the extra profits from the bargain.

Actual bargaining between networks and stations is much more complex. Under FCC regulations, stations are not required to accept the full schedule of the network with which they are affiliated. This rule severely limits the scope of affiliation agreements by prohibiting contracts of the form: station A receives X dollars of compensation if it transmits all of network N's programs.

The prohibition creates problems for the networks because their profits are tied to the size of the national audience of their shows. These revenues vary directly with audience size, while average costs per viewer for a given program vary inversely with size: the larger the audience, the larger the profits. Also, to the extent that viewers still prefer not to switch channels, adjacency effects exist among programs. A local station might elect to show a program appealing to a locally small but desirable audience, or one for which the local audience is larger than the network alternative. Although this decision may be profitable for the local station, it affects the network adversely. The small size of the local program's select audience means that there is a smaller lead-in audience to the following network show, and hence a smaller network audience and profits.

The FCC's prohibition of all-or-nothing (exclusive dealing) contracts between networks and affiliates forces the networks to use compensation schemes that encourage local affiliates to carry the network programs that are weaker or of limited local appeal. Although the networks compensate stations for most of the hours that the stations carry, they do not do so uniformly for all dayparts. (Besen and Soligo, 1973; Besen et al., 1984). Some network shows with great local appeal in a particular area will be especially profitable to affiliates because they can earn large revenues from the announcement time and spot commercials in those programs. If the profits on such shows exceed what the station can earn by using local or syndicated programming, then the station will carry them, with or without compensation.

In our one-network, two-station example, the network's optimal strategy is to identify such locally popular programs and then not compensate the affiliate for them. For the remaining programs the network need provide only adequate marginal compensation per program to induce the station to carry the network program. The amount will be just large enough to exeed the profits on locally originated or syndicated material. The network will provide this marginal inducement if its profits are increased by the process. If the necessary inducement is too large, the network will be forced to let the affiliate do its own programming. The process probably benefits the station. Instead of settling for a lump sum just sufficient to induce affiliation, the station receives the very popular shows free and can receive inducements for the other programs. In addition, because the FCC frowns on finely graduated compensation schedules, networks pay the compensation required to induce the clearance of marginal programs for some inframarginal shows as well.

The cases of one- and two-station markets are interesting because they reverse the relative power position of the local stations and the networks. Consider the one-station, two-network market. Because network average costs per viewer vary inversely with audience size, both networks want to clear as many shows as possible, as long as their profits from affiliation are positive. If a local affiliate of network A also affiliates with network B, A cannot credibly threaten to drop the station's affiliation—A would lose potential profits on its remaining programs by doing so. The station can bargain for the best shows of both networks and receive compensation from both. This example explains the viability of multiple affiliations in smaller markets. In sum, the networks must be responsive to the needs and interests of their affiliated stations.

Another way to characterize the network-affiliate relationship is to point to the "free rider" opportunity that exists when individual affiliates find it profitable to preempt individual network offerings in order to broadcast individual syndicated series that are locally more popular. The fact that network affiliation is, overall, more profitable for such stations than reliance on the syndication market for programs depends on the cooperation of other affiliates in sharing program costs and in accumulating national audiences of useful size. If each affiliate cleared only the individual network series that were more profitable than syndicated alternatives, the network would have to pay lower compensation, offer lower quality programming, or both.

As already noted, free riding is likely to make stations, networks, and viewers all worse off. It is likely to be a particular problem for the Fox network, which, because it has incomplete national coverage, can currently offer programs only slightly more profitable on average than the syndicated products they displace. Efforts by Fox to induce its affiliates to clear network fare must be even more vigorous than those of the major networks.

These free rider problems could be avoided in principle by vertical integration—outright ownership of stations by networks. But FCC rules prevent any entity from owning stations reaching more than 25 percent of all television households, and there would be economic disadvantages to outright ownership of all affiliates. Networks thus face some of the problems that exist for all franchise businesses, though with special force. The best solution from the point of view of the ultimate consumers—advertisers and viewers—involves leaving the network and the stations free to reach imaginative contractual solutions to their free rider problems. Given the highly competitive structure of the industry in the wake of the FCC's reforms, there is no reason for FCC regulations or antitrust restrictions to hamper this process.

Television News

Television news, a popular forum for vigorous political and social commentary, is also a business. Economic actors seek profit, and viewers seek satisfaction. The importance of this business to the networks is indisputable. CBS, not atypical, scheduled some 2,400 hours of news and public affairs broadcasts in 1990, almost 40 percent of its network schedule.

There are two kinds of television news broadcasts: local and network. Local news is locally produced and focuses on local news events, but it also generally includes national and international stories. Network news focuses on national and international events. Until 1980 there were three U.S. suppliers of television network news services to local television outlets: ABC, CBS, and NBC. The television network news services offered by these three networks include a nightly news broadcast of half-hour duration, coverage of special events of extraordinary interest such as space shots and political conventions, occasional public affairs broadcasts, and other regularly scheduled news broadcasts such as "CBS Morning News," "Nightwatch," and prime-time news briefs. In addition, the television broadcast networks separately supply affiliate news services that include excerpts from network news broadcasts and "news feeds," which are used by local television stations to produce more attractive local news broadcasts.

In 1980 the television news business was revolutionized when Turner Broadcasting System introduced Cable News Network, a

24-hour-per-day, satellite-delivered, network news and feature service marketed initially to local cable television systems. In 1983 CNN in turn introduced a second 24-hour service, first called CNN-2 and later CNN-Headline News (CNN-HN or HLN), which includes continuous half-hour television network news segments similar in format to those offered by ABC, CBS, and NBC. Later CNN offered television stations the right to excerpt hard news and features from CNN and/or CNN-HN for use in local news broadcasts. By the early 1990s CNN-HN was used by cable television systems and by television stations throughout the United States, and was also available in Europe, Japan, China, and the Soviet Union. Whittemore (1990) provides an insider view of CNN's early history.

Television stations and cable systems pay for news services in various ways. In the case of ABC, CBS, and NBC, some news services are "sold" to affiliated stations in conjunction with the network sale of station commercial time to advertisers. The net proceeds of these dual transactions are remitted by the network to the station in the form of compensation payments. (In addition, the local station usually sells "adjacency" time to advertisers on its own.) News feed and excerpts for inclusion in local news programs are sold separately by the three major networks to each station on the basis of a flat monthly fee that varies by market. CNN and CNN-HN sell time to national advertisers, and they also receive cash payments from television stations and cable systems.

Producers and distributors of network news must be in a position to cover fast-breaking events worldwide and to coordinate and edit the coverage of numerous simultaneous or near-simultaneous events. Producers require a large and well-financed operation that includes news bureaus, reporters, and camera crews around the world, ready access to international and domestic communications satellites, facilities for the rapid processing and editing of video tape and film, and a central organization of editors and producers to select and edit news stories. As little as 1 percent of the video tape and film exposed each day is ever aired. The results are presented to the viewing public through a variety of broadcast vehicles (regularly scheduled news reports, briefs, special reports, and so on).

In a competitive environment, the substantial fixed cost of news-gathering operations cannot be sustained without multiple broadcast or other vehicles to make the news available to a large audience and to generate sufficient advertising revenues. At the same time, the quality and timeliness of the broadcast vehicles cannot be maintained without the support of the news-gathering organization. Only CBS, NBC, ABC, and CNN can claim to have achieved the large news-gathering operation and broadcast vehicles needed to make national and international television news available to the public. The large investment required to maintain a news organization, to establish a number of different broadcast vehicles, and to arrange for local outlets over which to spread the fixed cost of the operation tends to inhibit entry.

The costs of maintaining these facilities and a large organization of skilled personnel are largely independent of the number of television outlets that carry the resulting news services. Moreover, the costs of the organization can be spread over a number of different news services. These characteristics of the news business give rise to economies of size.

Economies of size have aided CNN. Its relative wealth of outlets and vehicles makes it the closest competitor to the news divisions of the three major networks. CNN earned its first reported operating profit in the first quarter of 1985. Total losses through 1984 were estimated at $72.74 million. CNN-Headline News was added to the basic CNN operations at an annual incremental cost of about $15 million. It has grown rapidly since its inception in 1982.

But no other entrants have proved successful to date. After only 16 months in operation, the Satellite News Channel was terminated on October 17, 1983. The termination occurred when SNC was paid $25 million by the Turner Broadcasting System in partial settlement of an antitrust lawsuit. SNC, a 24-hour television network news service marketed to cable systems, was a joint venture of Westinghouse and ABC, two large and experienced media organizations. At the time it closed down, SNC reached 7.5 million households and had 350 to 400 employees. The SNC service was similar to that of CNN-HN, providing a "wheel" of news that was periodically updated and repeated continuously on a 24-hour basis. The settlement agreement included an undertaking by Westinghouse and ABC not to compete in offering 24-hour television network news services to cable systems for three years.[8]

SNC's experience illustrates the difficulties of starting and operating a complete service of television network news. SNC losses during its last year of operation are estimated to have been in the range of $35 to $40 million and total losses over its 16-month life to have been about $60 million. Losses were attributed to lack of advertiser support in the face of intense competition from CNN and the three broadcast networks.[9]

In response to SNC and CNN, CBS, NBC, and ABC expanded their own news broadcasts into the late night and early morning time periods. They added a total of 33 weekly hours of news in response to CNN's entry.[10] The commercial time available on national news broadcasts nearly doubled. In particular, the networks sharply increased the quantity and quality of news material available to their affiliates for insertion in local news broadcasts.[11]

CNN responded to the anticipated entry of SNC by introducing CNN-HN prior to SNC's launch. CNN-HN was offered free of charge to cable systems already taking CNN. Following SNC's actual entry, CNN lowered its carriage fees, sometimes drastically, to cable systems. After SNC ceased service CNN more than doubled its fees to cable system operators.[12]

Another factor in SNC's demise may have been its late start in establishing its own news-gathering organization and in attracting cable system operators to its service.[13] Experience plays a critical role in determining the effectiveness and quality of a product as complex as a national network news service. The two-year head start CNN had over SNC may have provided CNN with an important competitive advantage. Despite a budget comparable to CNN's, SNC was never able to achieve comparable acceptance by cable systems and therefore by advertisers.

Despite these apparent entry barriers, both the Fox Broadcasting Network and the Christian Science Monitor have announced plans to enter the network television news business, and there are various news services that provide national and international television news feed to local stations. These include the Europe Broadcasting Union (EBU) and VisNews. EBU exchanges news feed with the major broadcast networks as well. In addition to news from the major networks, stations may acquire national and sometimes international news through affiliation with a "cooperative" news organization that shares material developed by the member stations. INN,

CONUS, and the Newsfeed Network are the most prominent of these services.

Among the cooperative organizations, the INN product was the closest approximation to network news. Much of INN's news was originated by its affiliated stations and by the WPIX news organization in New York. International news was purchased. In its first year of operation, 1984, INN had a budget of $6 million. INN ceased operation in 1991. Both CONUS and the Newsfeed Network rely even more heavily on member stations for coverage of news.

There are economies associated with the development of news teams that specialize in the coverage of national news. The national news team develops expertise that aids in the interpretation of events. Furthermore, news of national interest is not a constant occurrence in the areas covered by most local stations. It is therefore not cost effective for local stations to maintain significant excess capacity that will be employed only periodically to cover news of national interest. More efficient is the practice, followed by the national networks, of relying on local news crews as a story breaks and sending in mobile network crews to follow the story as it develops.

Local stations in the 1980s greatly expanded their news programs, in some cases doubling the hours devoted to news. The stations found that they could generate larger or more attractive audiences more efficiently with news than with alternative syndicated programming, particularly on an incremental basis. Stations have always had some local news-gathering and broadcasting operations. These operations existed in part because of the FCC's traditional interest in news broadcasts as a means of serving the public interest. The expansion of local news broadcasts allowed the fixed costs of the local station's news effort to be spread over more hours. Similarly, in part owing to competition from CNN, the broadcast networks greatly increased news and public affairs service. CBS, for example, went from 770 hours in 1979 to 2,400 hours in 1990.

In evaluating their expansion plans, local station managers recognized that they already had at their disposal the news feeds from the networks. The news feeds provided abundant material on national and international events, including sports clips, of considerable interest to their viewers. The quantity and quality of video news feeds available to local affiliates had steadily increased, a process in which CNN played a central role. In addition to providing independent stations with access to a network news feed, CNN stimulated the other competing networks to improve their feeds. NBC implemented a major expansion and improved the quality of its news feed; ABC increased the quantity of news material in its feed by 35 to 40 percent.[14] The local stations were able to upgrade the quality of their news broadcasts, thereby improving the size and demographic character of their audiences and resulting in greater advertising revenues. In part because local television news has become a better substitute for the network news, CBS, ABC, and NBC news broadcasts have experienced declining audiences shares since about 1980.

In the late 1980s, the networks began to substitute news programs for entertainment programs. Although news programs draw smaller audiences than do entertainment programs, their cost is much less.[15] "Nightline," "60 Minutes," "48 Hours," "Prime Time Live," and other news programs have relatively low incremental costs because they build on the fixed-cost investment of the news division. Moreover, unlike entertainment shows, there is no regulatory ban on network production of news programs.

In sum, national and international television news is a profitable strategic business for broadcast networks. Fox's interest in providing news coverage is not surprising in light of all the networks' increased competition in markets that supply sports and entertainment. A news organization offers each network an opportunity to produce broadcast products in-house with relatively predictable, though modest, audiences and costs. Still, there is an obvious difficulty for other networks, such as Fox, that may seek to enter the business of gathering national and international news, as opposed to relying on news feeds purchased from others. The latter approach is taken, for example, by the cable business news services Financial News Network and Consumer News and Business Channel, each of which purchased most of their news coverage from Dow Jones, Reuters and the Associated Press. (FNN and CNBC merged in 1991.) With fewer outlets than the major networks and fewer vehicles than CNN, Fox or another newer network will have higher unit costs than its competitors. CNN's remarkable success demonstrates that the key to entry is to exploit the low incremental cost of additional broadcasts and to build on the high-fixed-cost base of a worldwide news-gathering organization. CNN's strategy of reselling its output repeatedly throughout the day, over two channels, and to cable and broadcaster customers at home and abroad, has been crucial to its success.

A review of the economics of television news is not complete without mention of one area in which news organizations present a problem for networks. The networks have difficulty in Washington in their periodic struggles with Hollywood suppliers over restrictive regulation, a story recounted later in this chapter, and with other economic interest groups, such as cable operators. These difficulties are attributable in significant part to political ill will created by the network news organizations. No news show can possibly provide coverage that is as extensive or as favorable as each politician would prefer, and it is no coincidence that the networks have long enjoyed a reputation in Washington for excessive power and arrogance. In contrast, the picture studios seem willing to provide friendly politicians with photo opportunities in the company of stars and to engage in other effective lobbying.

Spinning off the news organizations as independent commercial entities, so that the networks could purchase finished products from more than one, would go some way toward solving this political problem. But spinning off the news organizations would deprive the networks of the programming source with the most controllable costs and risks.

Regulation

The Federal Communications Commission forbids individual stations from entering into an affiliation contract with any network that fails to comply with FCC standards.[16] Because the networks supply most of the programs that are broadcast on affiliated stations, the FCC influences, both directly and indirectly, the networks' behavior.

The principal issue is network market power vis-à-vis stations and programmers. As we saw in Chapter 1, network dominance was the result of the FCC's own decision to restrict artificially the number of broadcast licenses.

During the late 1950s, the Commission's doctrine of "localism" in the control and origination of programming did not make economic sense. Affiliates found it more profitable to use network programs than local or syndicated series. Networks pressed affiliated stations to clear network programs, despite free rider problems. The Commission's response to the problem was to restrict the terms and conditions of affiliation contracts, with the aim of reducing the ability of station licensees to relinquish blanket control of programming to the networks.[17] The bargaining power of the networks, however, is based on their ability to affiliate with other stations in the same local market. This power, as already discussed, is related to the number of stations relative to the number of networks, and to the profitability of network programming. The FCC's restrictions on contract clauses do not significantly alter this power structure and hence have little effect on the behavior of networks and stations. Other developments also failed to promote localism. In the quiz show scandals during the 1950s, congressional investigators emphasized the indiscretions committed by advertisers in shows supplied to the networks. The FCC then encouraged the networks to take responsibility for program content, and thus reinforced a trend already begun for economic reasons.

In the middle and late 1960s, the FCC again faced the network power problem, this time in response to Hollywood program producers who regarded themselves as victims of network control.[18] The FCC's response was threefold: the financial interest rule, which prohibited the networks from acquiring equity and profit rights in programs produced by others; the syndication rule, which kept the networks from selling programs to television stations; and the prime-time access rule, which limited to three the number of hours per evening within prime time (7:00 to 11:00 P.M.) that could be programmed by the networks. Affiliates in the top 50 markets were barred from filling this time with network reruns. The purpose of the prime-time access rule, which became effective in 1975, was to give other program producers access to individual station broadcast time through the first-run program syndication market. (The syndication market had previously been limited, as a practical matter, to nonaffiliated stations and to periods other than prime time on affiliated stations.) The prime-time access rule led to a flowering of demand for what are known as "junk" programs—talk and game shows—that now dominate the "access period" between

Table 5.7 Percentage of total syndicated program audience

Type of program	1971	1989
First run	36	70
Off network	65	31

Source: Crandall, 1990f, Appendix A.

7:00 and 8:00 P.M. The demand for first-run syndicated programming outside the access period grew, however, with the growth of cable, domestic communication satellites, and the consequent increase in independent stations. Today first-run syndicated programs are far more popular than off-network reruns (see Table 5.7). The audience for all syndicated programs approximately doubled between 1971 and 1989. But the prime-time access rule prevents additional programming of the sort most viewers would watch if it were available.

The prime-time access rule thus runs counter to the economics of television program production and distribution (Crandall, 1971, 1972) and works against viewers' interests. The financial interest and syndication rules make even less sense. They do not promote competition or diversity, and they interfere with the efficient allocation of risk between program producers and networks (Noll and Owen, 1983; Besen et al., 1984; Fisher, 1985).

Risk Bearing

The efficiency of risk bearing in the television business has been restricted by the FCC's financial interest rule and by 1980 Department of Justice constraints on the terms of option contracts (limiting rights to order additional episodes of series at prices set in advance) and (now-expired) limitations on the networks' internal production of prime-time series. Although there is general agreement that the reallocation of risks from the networks to Hollywood is inefficient, there has been no successful effort to quantify the cost of the government constraints. Informally, most analysts have assumed the cost to be small. It may not be, however. Useful in considering this question are empirical results, discussed later in this section, from a study we made in 1985 of the relative "success" (as measured by length of run) of prime-time series programs from 1960 to 1985 for all three major networks.

Restricting the terms on which networks can purchase options to renew prime-time series has the same effect as restricting financial interests: risks are shifted so that the networks are forced away from their preferred risk-return position. The result is higher costs for initial network exhibition rights for prime-time series and lower network rates of return. The effects on network and Hollywood behavior may be substantial; the risk, as we shall see below, was adversely reduced about 25 percent in the period 1972 to 1979, after having increased 33 percent in the period 1963 to 1972. By 1980 the level of network schedule risk was approximately the same as in 1960.

The indications (based on extrapolation of pre–financial interest rule trends) are that network returns would be optimized today at substantially higher risk levels. To be fully validated, these results need to be confirmed with network cost data for the periods before and after the constraints were imposed. Unfortunately, sufficiently complete pre-rule data on program costs do not exist.

The sale of option terms, syndication rights, and other financial interests allocates the risks inherent in producing and broadcasting prime-time television series. Political and regulatory debates spanning the past two decades have focused on these rights. In short, Hollywood persuaded the government to change market outcomes. The effect has been to shift more of the risks of prime-time series away from the networks and toward Hollywood, where they appear to be borne less efficiently—that is, at higher cost to society and to the networks.[19] Competitive market outcomes generally are efficient; if not, the parties will alter their contracts in search of mutual gains from the elimination of avoidable costs. The producers' argument that buyers in the series market are not competitive has little weight.

A television series is an asset consisting of a bundle of broadcast rights. The value of these rights depends on the success of the series in attracting audiences. The values of initial network exhibition rights, the right to renew the network exhibition rights (options), the right to profit from the receipt of music royalties, the right to earn revenues from syndication of a successful series, and the rights to other streams of revenue are all interrelated.

The ability of a series to attract audiences is not perfectly predictable; significant sums of money must be invested before it is clear how much profit, if any, will be earned by the owners of the broadcast rights. Hence all of the broadcast rights in each series are, and remain for some time even after initial broadcast, risky. The ownership of rights that have value only if the series is successful is a risk-bearing function; such risk bearers finance initial production. Not all rights are equally risky. The owner of the initial network exhibition rights may earn a reasonable return with relative certainty; the owner of renewal options or syndication rights in the same series may be more likely to lose his or her entire investment. In general, the risk associated with ownership of initial network exhibition rights is less than that associated with the other major broadcast rights.[20]

Suppose that small Hollywood producers are less able or less willing to bear risk than are the large studios. Leaving the networks aside, in a free market the smaller producers will be able to sell profitably rights that transfer risk to the studios, lowering overall production costs. Accordingly, if large Hollywood producers are less able or less willing to bear risk than the networks, the risks are most efficiently shifted to the networks. A rule that prevents transfer of these risks creates inefficiencies that increase costs.

When the various federal restrictions were imposed, two immediate effects and one longer-term effect could be anticipated as a result of the shift in risks. First, network payments to Hollywood should have been lowered, simply because fewer rights were being purchased. Second, the price of initial network exhibition rights should have increased because the discounted expected value of the revenues from the other broadcast rights would not be sufficient to induce Hollywood to produce the same series. Third, the prices paid by the networks for subsequent-year renewals of network exhibition rights to successful series should rise in the long run, reflecting the fact that the risks associated with such successes were being borne by the producers, who therefore would earn the rewards of success. Other factors, such as inflation, changing program standards, and growth in advertiser demand, would also affect the prices paid for programs.

An example may help to explain why options transfer risk. Suppose a series is either a success or a failure and that this becomes known during the first year it is on the air. If a series is not successful, it has no further value. If it is successful, then we assume it will earn predictable profits in future years. The likelihood of future success is much more assured for a series that has been successful in the past than for a new series in its first network season. This means that the profits are much easier to predict (more assured, less risky) after a series has proved to be successful than before the initial network contract is signed.

Consider a schedule of *network* profits from the broadcast of a hypothetical typical series (see Table 5.8). Remember that in this simple case, a series is either a "success" or a "failure." This example compares a risky asset with a riskless asset, each with the same expected return. The risky asset is the right (the option) to renew the series at a specified price. A sum of cash equal to the expected value of the return on the risky asset (the option) is the equivalent riskless asset.

If the network buys a four-year option contract with the initial broadcast rights for this series, it will be willing to pay up to $210, based on the $1,200 profit that comes with a 10 percent probability (expected value $120) and the $100 profit that comes with 90 percent probability (expected value $90). (A program that produces only a $100 profit is deemed a failure and will not be renewed; other, untried, programs will have higher expected returns.) These calculations leave aside time and risk discounts.

A three-year option contract is worth only $180 to the network. In the first year of the contract, if the series is successful, the producer will be able to negotiate a payment of up to $300 for the fourth year when the contract is extended. (How much of the $300 value to the network gets captured by the producer depends on the factors affecting bargaining power.)

Table 5.8 Network profits: a hypothetical case

Year	Success (10% chance)	Failure (90% chance)
1	300	100
2	300	0
3	300	0
4	300	0

Producers thus prefer short option periods if they prefer risk. Short options force producers to bear a greater share of the production costs in return for a "lottery ticket" that pays off if and only if the series is successful. The option period in this example is nothing more than a device to shift risk from series producers to networks. The longer the option period, the more risk is borne by the network, and the lower will be the contractual, per episode payments during any given year of the option period. The shorter the option period, the more risk is borne by the supplier and the higher will be network payments per episode for successful series in any given year of the option period.

Networks are hurt by rules that reduce option periods below marketplace levels if they can bear risk more cheaply than can producers or if they can, through scheduling control, affect the risks. Although it is very difficult to measure either of these effects, their significance probably increases as the option period grows shorter. An effect, perhaps small to begin with, becomes insignificant when discounted six or eight years into the future. The same effect may be palpable when realized only two or three years hence.

Of course in the real world no series, regardless of its popularity in a preceding year, has a 100 percent likelihood of success in the following year. Our study of network prime-time series renewals demonstrates, however, that new series have a low probability of success and that series that have been renewed have a much higher probability of success. We computed the average frequency with which series premiering between 1960 and 1981 were renewed following one to four years of network exhibition (see Table 5.9). Averages were computed separately for 1960–1971 and 1972–1981

Table 5.9 Probability of renewal for prime-time series

Number of years on network	1960–1971	1972–1981
1	36%	24%
2	63%	72%
3	72%	84%
4	70%	76%

Source: Computed by authors from published network schedules for television seasons from 1960–1961 to 1981–1982.

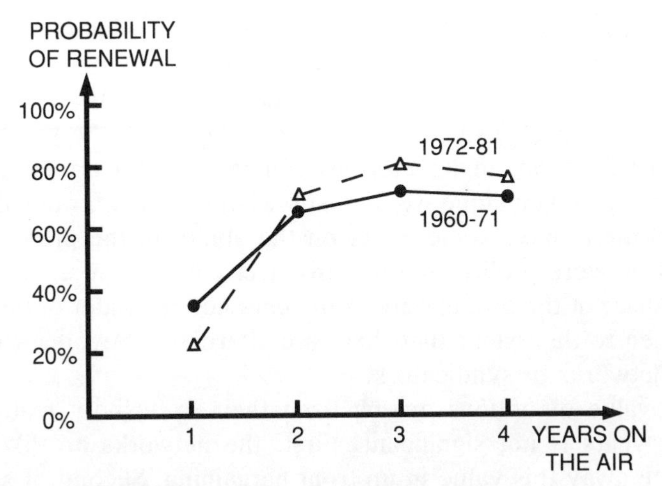

Figure 5.2 Probability of renewal of prime-time series

to allow for possible changes in this relationship due to the imposition of the financial interest and syndication rules in the early seventies. In both subperiods the likelihood of renewal following the first year of exhibition is much lower than for series that have already been renewed at least once. The likelihood of renewal following the initial network year averaged 36 percent from 1960 to 1971 and 24 percent from 1972 through 1981. In both subperiods the likelihood of subsequent renewal for a series renewed the previous year is double to triple the likelihood for first-year series (see Figure 5.2).

The renewal study suggests that a year of exhibition provides a network with information that it cannot acquire through other means regarding the probable success of a series. Networks are much more successful in predicting the future performance of series with a track record than new series. For this reason renewal decisions are much less risky than initial purchase decisions.

The option years in the initial contract constrain the ability of the producer to raise the price. The producer can really bargain only for the first year beyond the end of the contract. The fact that producers initiate renewed negotiations suggests that the networks are better risk bearers. Annual renegotiations of the current contract price reflect the moral hazard problem: producers and talent can find subtle ways not to perform so that they can force the network to share with them some of the profits from a successful series. The network for its part has to behave consistently with the expectations it has created in Hollywood (Klein, 1980). Presumably this fact is reflected in the initial bargaining, with the result that unsuccessful series make even less for their producers than their performance taken alone would justify. However, contractual obligations surely have some effect on the ability of the producer to extract network profits, and the producer can hope to capture the lion's share of the profits only from years not yet under option and only then to the extent that there are alternative buyers, whether other networks or syndicators.

The value of options arising from their exclusivity feature in distant years is not significant. First, the networks are likely to compete away this value in up-front bargaining. Second, if scheduling strategies and "audience flow" surrounding successful series are still as meaningful as one is led to believe, it is doubtful that another network will place a comparable value on the series. Network programmers continue to believe that the size and type of audience for a given program are heavily influenced by preceding programs on that channel.

Until 1980, when the Department of Justice changed the rules, longer option periods were freely negotiated, which suggests that there were net gains to at least one and more probably to both parties from network risk bearing. Changed circumstances may have shifted the optimal locus of risk bearing to some degree. But the fact that today the parties negotiate four-year options similarly suggests that moving to three-year options as a result of an extramarket constraint would reduce efficiency and therefore reduce potential profits for the networks.

It appears that networks pay more per episode for successful series with short option periods than for those with long option periods, but this prediction cannot be tested directly. The Department of Justice consent decrees did not reduce option periods below then-existing marketplace levels. To examine the likely effects of reducing option terms, we must examine instead the effects of the financial interest rule imposed by the FCC in the early 1970s. The effects of this rule ought to be the same, in direction at least, as the effects of shorter options. The financial interest rule changed marketplace operations, and therefore its effect ought to be measurable.

A comparison of invoices may help to illustrate how the financial interest rule affected network payments to Hollywood (see Table 5.10). Consider hypothetical—but representative—Invoice A from a Hollywood studio to a network for the rights to a new series in 1970; Invoice B is another hypothetical invoice for the same series if it were purchased new in 1975, several years after the financial interest restrictions were enacted. The invoices itemize the prices of the various rights purchased in order to show the effects of the FCC rules. In practice it is the price for the package as a whole that is negotiated, rather than each element. Nevertheless, putting prices on the elements of the package helps to illustrate the effects of the rules.

Table 5.10 Two hypothetical invoices from ACME Studios

Invoice A (1970)

For: Rights as itemized to *A Prime-Time Series,* to be first broadcast in September 1970. This invoice covers one year's worth of episodes and reruns, including *22* original episodes and up to *18* reruns.

Item	*Per Episode*
Initial network exhibition rights	$60,000
Same-year reruns (up to 2 per episode)	$20,000
Renewal option (8 years at above prices with escalator)	$25,000
Right to 50% of profits from syndication	$25,000
Package price, total (1970 dollars)	$130,000
Total package, 22 original episodes:	*$2,860,000*

Invoice B (1975)

For: Rights as itemized to *A Prime-Time Series,* to be first broadcast in September 1975. This invoice covers one year's worth of episodes and reruns, including *18* original episodes and up to *18* reruns.

Item (in 1970 dollars, unless otherwise indicated)	*Per Episode*
Initial network exhibition rights	$80,000
Same-year reruns (up to 2 per episode)	$20,000
Renewal option (8 years at above prices with escalator)	$25,000
50% financial interest in syndication profits	*not permitted*
Package price, total before inflation	$125,000
Inflation adjustment (50%)	$62,500
Package price, total (1975 dollars)	$187,500
Total package, 18 original episodes, 1975 dollars:	*$3,375,000*

Note that in Invoice B (1975) the price per original episode of initial network exhibition rights has increased from $60,000 to $80,000 in 1970 dollars. Such an increase (no one knows the exact amount) is the predicted effect of the FCC restriction preventing the networks from acquiring the right to profit from syndication. In effect, what has happened is that the network was willing to pay $25,000 for a 50 percent interest in the risky syndication profits, whereas the studio valued those rights at only $5,000. Therefore, forced by the federal rule to retain for itself the financial interests formerly acquired by the network, the studio requires an extra $20,000 for the initial network exhibition rights in order to induce it to produce the same episodes as invoiced in 1970 and accept the risk associated with an increased financial interest. The $20,000 is the "inefficiency" caused by the rule. The invoices also illustrate some (not all) of the other factors that are difficult to keep equal. For example, fewer original episodes are ordered in 1975 than in 1970, and there has been general inflation.

The effects of the rules on the network can also be discussed from a portfolio theory perspective. In constructing a portfolio of investment assets, an investor will choose among a variety of investment opportunities that vary in riskiness and in expected return. It is generally assumed that investors are risk averse. This means that in selecting among assets with the same expected return, an investor will choose the asset which is least risky. Similarly, for assets of equivalent riskiness, an investor will choose the investment with the highest expected return. An investor can adjust the risk and expected return of a portfolio by changing the mix of assets in the portfolio. The "efficient asset frontier" is the set of portfolios for which it is not possible either to increase the overall rate of return except by increasing the risk, or to decrease the risk without decreasing the rate of return. In other words, for every portfolio inside the efficient asset frontier, there are one or more achievable positions on the frontier that the investor prefers. In contrast, movements along the frontier require a trade-off between risk and return.

The pre-rule efficient asset frontier is PP′ in Figure 5.3. The network optimizes its risk-return position by choosing a portfolio of broadcast rights on this frontier and then further leveraging its position by borrowing to get to point A. The set of portfolios among

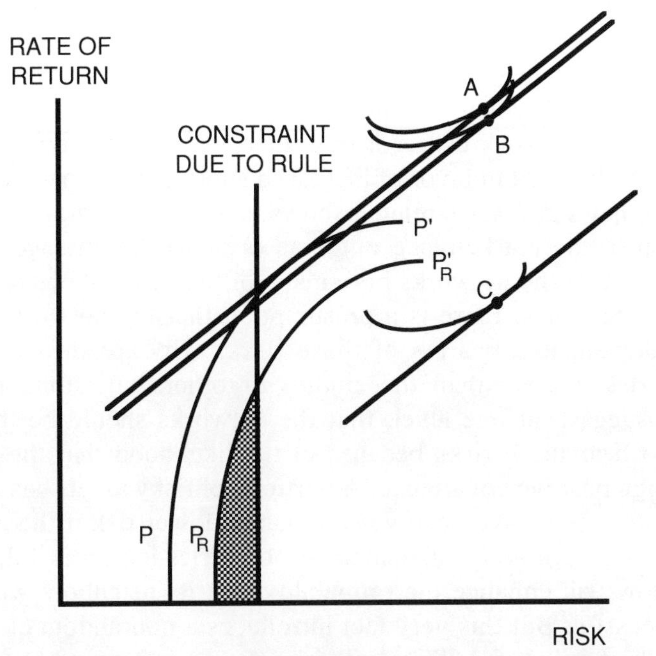

Figure 5.3 Effect of rule on risk allocation

which the network can choose is limited by the financial interest rule to those to the left of the vertical line. The network is first forced from its preferred leveraged position at A to a constrained position at B. As this happens, its rate of return must decline (B lies below A).

Those assets the network is now permitted to buy (that is, initial network exhibition rights) will be priced higher than before, reflecting the inefficiency of risk bearing by Hollywood suppliers. The entire remaining opportunity set of rights portfolios is therefore shifted down to $P_R P'_R$. Assets with the same risks now offer a lower expected return and assets with the same expected return reflect higher risks. Simply stated, Hollywood increases the prices of initial broadcast rights. The assets available to the network, taking account of the direct and indirect effects of the rule, are indicated by the shaded area in Figure 5.3. The new, optimally leveraged portfolio available to the networks necessarily entails an overall lower rate of return. This new position is represented by point C.[21]

Faced with higher prices for initial exhibition rights to otherwise comparable prime-time series, the networks naturally substitute other kinds of programs. This effect is illustrated in the sample invoices by the reduced number of original episodes ordered. But in addition, network usage of traditional prime-time series during the 1970s declined in favor of increased sports, made-for-television movies, news and news-related shows, and other program forms.[22]

Because the marketplace outcome prior to the passage of the rules was that the networks bore many of the risks associated with prime-time series, there is a presumption that the networks were the more efficient bearers of those risks. Why are the networks better risk bearers than the studios or others? Portfolio theory would suggest, at one level, that the networks should be the last ones to bear these risks because of the likelihood that the series have high positive covariance. A portfolio of risky assets has higher risk if its assets have positive covariance, lower risk if the assets' returns are negatively correlated. If one series is successful, audience flow will enhance the probability that its neighbors will also be successful. But this very fact introduces a nonrandom element. The risk of failure faced by any series can be influenced by scheduling decisions. Thus there are potential gains from internalizing these externalities among programs in the schedule.

The analysis of the effects of the ban on networks' engaging in syndication of programs they air is similar. If the networks engaged in syndication, or at least if they bore the risk of syndication, they would not need to make "bulk sales" of the syndication rights in their in-house productions. When these rights are auctioned off, the networks' risk-free receipts are lower than would be the (expected) risky receipts from engaging directly in the syndication of such properties. If the networks are able to bear the risks of syndication more efficiently than are independent syndicators, the effect is to drive down the networks' risk-adjusted return. Society is worse off because of the waste of resources in using a more expensive provider of service.

Taking the networks out of the risk-bearing business makes them less concerned about the effects of a decision about the scheduling of series B on the long-term profitability of syndication rights and other downstream interests in series A. This issue is what the debate concerning financial interests was all about: that the networks could and would take account of these interactions among the various program rights if they had an interest in the resulting profits or losses, leading to a "bias" in favor of series in which the networks had an ownership interest. From the standpoint of economic efficiency, consumer welfare, and business, it would be entirely proper to have such a bias. From the Hollywood producers' point of view, such "favoritism" seems unfair, and to them it is no answer to say that favoritism promotes economic efficiency and competitive markets.

Suppose that a program has been purchased for network exhibition. The producer retains the syndication rights. In deciding where to place that program in the schedule, the network considers the effects of placement on the overall profitability of its schedule; it need not consider the effects of its scheduling decision on the likelihood that the show will be renewed for another season and hence on the value of the syndication rights to the show. The network's indifference to this important issue would lead the producer, if it could, to put into its contract with the network some inducement to behave more efficiently. The difficulty of doing so would lead, absent the FCC rule, to the network's acquisition of a direct interest in the rights. This interest would provide an incentive toward behavior that would maximize the overall value of the show to society as a whole.

In the public debate concerning the imposition of the restrictions, the networks argued that any favoritism by the networks toward series in which they had a financial interest would be offset by the short-run necessity to maximize net advertising revenue. Indeed, these short-run effects may dominate, but the long-run effects may be much more significant than has been thought.

One reason for the difficulty of measuring the economic effects of risk shifting is that the losses caused by inefficient risk shifting (the $20,000 in the sample invoices) are likely to be swamped by the other effects of the federal rules. Because the networks' contracts with Hollywood are so complicated and multidimensional, many have assumed that the networks will somehow find other means to compensate for the constraints imposed by government regulations. For example, option terms may have lengthened. If so, there would be little if any effect on the prices of initial broadcast rights, making measurement even more difficult.

If as a result of the regulations the networks pay too much for the broadcast rights they do buy, one would expect to see effects on the riskiness of program schedules. One possibility is that both Hollywood and the networks would prefer less risky program strategies. The networks, if deprived of the ability to bear risk and to earn downstream profits from the exploitation of successful properties, and paying too much for broadcast rights of given risk because of the need to induce Hollywood suppliers to bear those risks, would tend to shift toward less risky strategies in order to lower costs. As a result, one would expect to see fewer experimental concepts, fewer deals with unknown producers and talent, and a narrower range of dispersion in the average length of runs for series (that is, fewer wild succȩsses). Similarly, the Hollywood producers, enjoying higher prices for initial network exhibition rights that would nevertheless be insufficient compensation for the risks now borne by them, would be less willing to offer series with risky formulas. For both reasons, the riskiness of network series should have decreased as a result of the financial interest rule. How much it decreased would be a rough and indirect measure of the adverse impact of the rule on the networks' costs.

If it were true, by contrast, either that the inefficiency effect of higher-cost Hollywood risk bearing were quantitatively insignificant or that other facets of the contractual relationship between the network and the series producer provided an escape valve, then one would expect to see little effect on series and schedule risk patterns.

To estimate the degree to which the riskiness of new prime-time series was affected by the financial interest rule, we compiled from published network schedules data on the length of run for prime-time series and on ratings on all three networks from 1963 to the early 1980s, spanning the decades before and after the rule went into effect.[23] The riskiness in the prime-time schedules of each season was measured by using the variance of the remaining length of run of all the series broadcast that season and by looking at the variance in ratings among shows in each season. A prime-time schedule made up of series, each of which lasts two seasons, will show far less variance (less risk) than a schedule whose series last two seasons on average (some last two months and others last five years). This is analogous to any asset portfolio. A stock portfolio that contains unrelated assets of widely variant quality, but whose average or expected yield is 10 percent, is riskier than a portfolio of unrelated assets of equal quality, each with a 10 percent expected yield. Length of run appears to be correlated with the popularity of the series and highly correlated with the rate of return to the owners of rights in the series.

The data on length of run for prime-time series reveal a rather striking pattern. From the early 1960s to about 1972, the riskiness of television program schedules, as measured by their variance, increased (see Table 5.11).[24] Between 1963 and 1972, for example, the riskiness of prime-time series increased by 33 percent. If the 1963 to 1972 trend were extrapolated, risk levels would have been

Table 5.11 Prime-time series: number, length of run, variance (shows premiering after 1959)

	All series			New series		
Season	Count	Average	Variance	Count	Average	Variance
1963	70	2.31	5.21	42	1.52	2.39
1964	81	2.16	4.32	41	1.70	3.04
1965	86	2.27	4.66	42	2.10	4.70
1966	86	2.22	3.88	46	1.50	2.39
1967	79	2.47	4.60	34	2.18	6.45
1968	83	2.39	5.06	34	2.13	5.96
1969	77	2.39	4.64	27	1.96	3.63
1970	78	2.19	5.76	30	1.64	6.66
1971	62	2.44	5.78	30	1.60	3.80
1972	62	2.70	6.82	28	2.34	8.39
1973	67	2.36	6.52	30	1.42	4.86
1974	73	2.42	6.87	39	1.71	6.20
1975	87	2.21	6.11	45	1.25	3.34
1976	91	2.02	5.44	48	1.27	3.91
1977	92	2.05	6.38	49	1.33	6.08
1978	105	1.97	5.80	63	0.99	2.55
1979	95	2.12	5.88	47	1.42	5.70
1980	83	2.30	5.81	37	1.40	4.79
1981	86	2.21	5.60	41	1.45	4.50
1982	88	2.23	5.58	45	1.55	4.47
1983	90	2.12	4.58	47	1.24	2.64

Source: Computed by authors from published network schedules for indicated television seasons.

Note: Programs still on the air as of January 31, 1990, were assumed to run until September 1, 1990.

substantially higher in 1980 than in 1972. But starting about 1972 riskiness began to decline. By 1979 risk had decreased by about 25 percent from its 1974 peak. The pre-rule trend and the post-rule experience may be compared graphically (see Figure 5.4). Ratings tell a similar story (see Table 5.12).

The same data were also summarized with respect to the riskiness of series that were *new* in each season (see Table 5.11). The data for all series, as opposed to just new series, are probably better indicators of overall schedule risk, but the basic trends are the same in any case. It was around 1972 that the effects of the FCC's financial interest rule first became apparent to buyers and sellers of prime-time series broadcast rights.

These risk effects are far larger than expected. In the 1970s, of course, other factors increased risks: changes in the competitive structure of the business and the prime-time access rule, which decreased the size of the networks' program portfolios. The risk *reduction* associated with the financial interest rule was therefore much more dramatic than the net figures indicate.[25]

The data suggest one economic explanation for the long political struggle between Hollywood and the networks. In the mid- to late 1960s there was a steady increase in the riskiness of the network schedules, presumably associated with greater use of small independent producers and lesser-known talent. Indeed, in the FCC's 1983 review of the financial interest and syndication rules, the studios' consultant, ICF, documented increased concentration of production and a decline of independent producers after the rules were adopted.[26] Network behavior in the late 1960s may have threatened the *internal* stability of the Hollywood establishment, triggering the political effort orchestrated by the major studios. In this light, the federal restrictions can be attributed as much to internecine Hollywood discord as to a struggle between Hollywood and the networks.

Table 5.12 Ratings variance

Period	Mean rating	Mean variance
1963–1972	18.15	22.94
1973–1983	17.29	17.66

Source: Crandall, 1990f, Appendix D, p. 5.

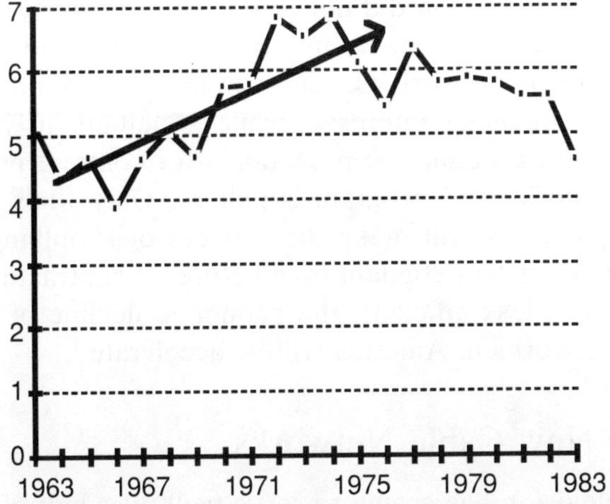

Figure 5.4 All prime-time series: seasonal variances of remaining length of run

The federal restrictions have not reduced the tension between Hollywood and the networks. The studios used political leverage to react to perceived threats to their welfare, but they were unable to control their own or the networks' competitive market reactions to the new structural conditions created by the rules. Faced with greater risk as a result of their own efforts, but unwilling or unable to bear that risk at prevailing prices, the studios increased prices and changed the programming they offered. Worse, from the networks' point of view, is the fact that the regulations increased concentration among Hollywood sellers and reduced the effectiveness of a profitable program form—the classic prime-time series—at a time when the networks faced more competition for audiences from existing products exploiting new windows.[27] The major studios gained from this activity at the expense of smaller independent producers and the networks.

In 1990–1991 the financial interest and syndication rules were debated once again. As in 1983, the Hollywood studios commissioned an economic report with a new justification for the rules, and they—along with the networks—pulled political strings. The three major broadcast networks, now supported by Fox, fought back.[28] But despite the lack of substantive merit in the rules, and despite the fact that the White House, the Justice Department and the Federal Trade Commission supported repeal, the Federal Communications Commission did not repeal them. Bitterly divided on the merits, the commissioners in April 1991 substantially relaxed the restrictions on network participation in foreign syndication and ownership of financial interests, while simultaneously imposing new regulations affecting the procedures for contract negotiations between networks and program suppliers. Only time will tell if the new regulations will make the process of supplying network programs more or less efficient than before. If contracting for programs becomes less efficient, the economic decline of the major broadcast networks in America will be accelerated.

The New Cable Networks

As cable television has grown to serve well over half of all television households in America, there has been an explosion of new cable television networks. By 1990 there were about seventy networks serving cable subscribers, with announcements of a dozen more in the pipeline. Most of these networks (which are analyzed more fully in the next chapter in connection with their ownership by cable operators) rely on advertiser support as well as payments from cable systems.

The new networks are generally specialized, although a few, such as Turner Network Television (TNT), the Family Channel, and Lifetime, seem to ape the general entertainment formats of the major broadcast networks. Entry and exit are so free and easy that competition is very intense. Even those cable networks that are the only occupants of particular format niches, such as the Weather Channel or the Travel Channel, do not seem to be in a position to exercise market power over cable operators or advertisers. It is noteworthy that most cable networks rely on programming that is produced for a number of other windows or whose costs are otherwise shared with additional media vehicles. The phenomenon of windowing, discussed in Chapter 2, is central to understanding why the supply of cable networks is so elastic.

The Decline of the Broadcast Networks

Until the early 1970s the networks were protected from external competition in national television advertising markets and in the business of supplying national programming to local stations. The protection arose from the FCC's initial failure to allocate sufficient VHF spectrum to television broadcasting, and it was perpetuated by the successful efforts of local stations and networks to prevent FCC actions (such as deintermixture of VHF stations and UHF stations in the same markets or deregulation of cable television) that would increase competition. This period of FCC protection also was marked by increasing demand for national audiences by advertisers, and by high and growing profits for networks and for VHF network-affiliated broadcast licensees, especially the stations owned and operated by the networks.

Then the economic and political foundations of the broadcast networks' protected position began to erode. In the early 1970s the FCC's open skies policy for domestic communication satellites led to competitive entry in the satellite business. Soon the cost of national video distribution to broadcast and cable outlets began to drop. In 1971 the FCC for the last time imposed new limitations on cable television importation of distant broadcast signals. The FCC also began to license more independent stations, most of which were UHF. In 1972 the FCC restricted the networks' ability to acquire financial interests in prime-time entertainment programs, and it forced divestiture of the networks' syndication businesses. In 1974 the Justice Department brought its antitrust lawsuit against the networks, which ended in consent decrees restricting in-house production of prime-time series entertainment programs and option terms. In 1975 the FCC's prime-time access rule, which may have increased competition in the network business by strengthening ABC, became effective.[29] Satellites and cable soon encouraged the development of independent stations and of first-run syndication as a source of competition for the networks.

During the second half of the decade, the FCC's restrictions on pay television were overturned by the courts, most of the remaining restrictions on cable television were eliminated, and the Justice Department successfully attacked the NAB Code restrictions on commercial units. By 1980 explosive growth in VCR penetration had begun. About the same time, the FCC opened the door to new video-delivery technologies, including direct broadcast satellites, low-power television stations, and multipoint multichannel distribution systems.

As a result, the three broadcast networks entered the mid-1980s much weakened. The artificial scarcity of spectrum that had been the original source of barriers to entry was no longer effective in protecting the networks from competition. The broadcast networks had been a bottleneck between program producers and national advertisers, on the one hand, and local stations and audiences, on the other. But regulatory reform and new technologies in the 1980s permitted producers and advertisers to bypass the networks. Consequently the networks' share of the viewing audience began to decline, their unit costs began to rise, and their advertising products became less unique. Given the networks' inability to adjust their cost structures quickly to the new realities, only continued growth in demand by advertisers kept disaster at bay.

Despite the networks' loss of market power, federal controls, based on the theory that the networks had excessive power and influence, continued to handicap the networks in competing with their new rivals. The ability of the networks to control events at the FCC and in Congress deteriorated further. In battles over network deregulation, cable and Hollywood interests generally succeeded at the expense of the networks in spite of an intellectual climate sympathetic to the networks' position. One source of the networks' political weakness was the continued perception of arrogance and power in the networks' news divisions.

The three major broadcast networks' share of the television audience declined from over 90 percent in 1970 to less than 60 percent in 1990. The networks' share of national television advertising declined to less than 50 percent. Similarly, the three-network share of total expenditure on television programming (including television rights to movies) also dropped to less than 50 percent. Fierce competition for viewers and advertisers developed from national cable television networks and from independent stations airing syndicated products. One competitor, Fox, successfully entered the business of providing conventional broadcast network services. By 1987 ratings for the ABC network (measuring its percentage reach of all television households) had declined to levels not experienced since the 1950s, when ABC's affiliates did not cover most of the country. Figure 5.5 shows the three-network prime-time rating and share for the seasons 1968–1969 through 1988–1989. (Ratings are

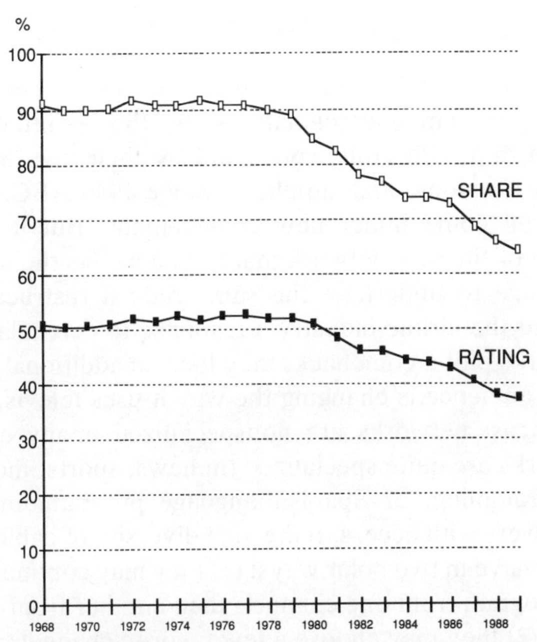

Figure 5.5 Three-network rating and share of prime-time audiences, 1968–1989

expressed as a percentage of all households with televisions; share is expressed as a percentage of households actually viewing television.)

The business managers of the broadcast networks have been hit hard by declining advertising and viewer shares. Hundreds of network employees have been fired or transferred in corporate slimdowns. Networks' formerly stable relationships with affiliates have changed. In the 1990s networks drive harder bargains and in some cases have proposed to slice annual compensation to affiliates for carrying shows by as much as 20 percent.[30] Hollywood, with many new channels and customers clamoring for its products, has not reduced its charges to the networks for programming: the average per hour cost (in nominal dollars) of a fall series more than doubled from 1978 to 1989. The networks are casting about for recovery strategies more imaginative than cost cutting; NBC is diversifying into direct broadcast satellite systems and cable networks. CBS is buying sports rights in large quantities.[31]

The Future of the Broadcast Networks

The broadcast networks face an uncertain future. They are constrained by governmental regulations, by the resulting limits on their relationship with Hollywood, and by their own interactions with affiliates, unions, and suppliers. Since 1986 ABC, CBS, and NBC have all come under new management. But it is unclear whether any of the new network managements has the imagination or the courage to undertake the same radical restructuring that characterized the airline industry's response to deregulation.

As networks plot a comeback, they face an additional challenge: the viewing audience is changing the way it uses television. Traditional broadcast networks are nonspecialized; many of the new cable networks are quite specialized (in news, sports, movies, children's programming, or Spanish-language programming, for example). Viewers with access to the vast diversity of cable programming may behave in two polar ways: (1) they may continually graze, picking a program from one channel, then another from a different channel, or (2) they may choose a few favorite channels and watch those, seldom trying other channels. Early evidence on viewing behavior in cable households suggests that the second of these patterns is more typical, just as it was typical for viewers of the traditional broadcast networks to keep watching the same channel until provoked. This pattern of behavior formerly led to an audience flow through a time stream of centrist, modal-taste programming; in the 1990s it leads to fragmentation.

Today cable is available to nine-tenths of the nation. Half of all viewers have access to 30 or more channels. Two-thirds of all television households have VCRs and three-quarters have remote controls. Most viewers are in a position to design a personal viewing diet that is entirely different from that of their neighbors. If viewers become attached to the custom offerings of cable, and to particular channels, it will be increasingly difficult for networks to make modally focused programs attractive. James G. Webster (1986, p. 77) notes: "Two of the principal features of the new media—the correlation of content with channels and the differential availability of those channels—produce a degree of audience polarization unprecedented in our experience with traditional forms of television."

If specialized cable channels do not begin to compete by becoming less specialized (a possibility that depends on their ability to attract significant ad revenue by doing so), then the relatively eclectic, general-interest format of the networks may not remain viable. Lingering regulatory constraints, traditional intra-industry relationships, and simple rigidity may doom the traditional networks to a long painful decline, akin to that of *Life* magazine a generation ago, as the middle ground of last resort.

Most of the networks' new competitors can rely on a combination of advertiser support and direct and indirect viewer payment. The broadcast networks are confined by technology, and perhaps by political reality, to advertising support. They may be unable to overcome this handicap, the effect of which is to permit nonbroadcast networks with much smaller audiences to compete effectively for network quality programs. FCC rules, designed (however unsuccessfully) to contain the power of the networks in the days when they were powerful, prevent the networks from being efficient competitors today. What, then, are the networks' strategic options in the 1990s?

Strategic Options

Although the networks have various strategic options, many cannot be undertaken without some change in the regulatory and political framework. Virtually none of the FCC restrictions on network business activities can be justified by substantive public policy analysis; the restrictions are perpetuated because they serve the economic interests of politically strong groups, such as the Hollywood studios, or because they preserve a political raison d'être for regulators and congressional subcommittees. One issue that the networks must address is which regulatory or political battles to take on in the years ahead. An understanding of the economics of each strategic option must thus include consideration of the costs and benefits of maintaining each of the current restrictions on network activity. Almost any strategic option affected by these restrictions that is profitable for the broadcast networks is likely to be more profitable for their cable network competitors, because the latter lack regulatory constraints and therefore can act more quickly on each opportunity.

Harvest

The harvest strategy involves charging relatively high prices, failing to invest in the business, and letting market share drift downward while maximizing cash flow. It assumes that broadcast networks will be producing a commodity product with low levels of additional investment and risk. No longer will the three networks or any competitor be able to produce a unique product: access to nearly 100 percent of American households. For some period the three networks will continue, however, to produce the largest *available* national audiences, because competing national networks do not as yet have access to as many television households. Investment in new technologies and new services is unwise, because the absence of federal restrictions on the networks' competitors allows faster reactions and quick imitation. Profitable operation calls for minimization of operating costs, no significant new investment, and pricing to yield a gradually declining market share. Opportunities for innovation arise chiefly in finding new ways to lower costs, most likely in program forms, and also in compensation payments to affiliates. Eventually, following this strategy in the face of continued restrictive regulation, the broadcast networks may end up as the passenger railroads did.

Backward Vertical Integration

The strategy of backward vertical integration involves the acquisition of control (through merger or internal development) of program production activities. Such a strategy will be profitable if there are economies (cost savings) from vertical integration, or if control of the supply of network programs will have an exclusionary or cost-raising effect on rivals. Backward integration includes the full acquisition of property rights, including options to renew network broadcast rights and financial interests in syndication and other ancillary revenues, in programs produced by others. These tactics are heavily constrained by FCC and Department of Justice regulations, which also would probably restrain any of the three major networks from acquiring or being acquired by a Hollywood studio to compete more effectively with the new Fox network.

Forward Vertical Integration

The strategy of forward vertical integration involves the acquisition of control (through merger or internal development) of retail distribution outlets, such as broadcast stations, cable systems, or video cassette rental stores. Such a strategy will be profitable if there are cost savings from vertical integration, or if control of distribution outlets will have an exclusionary or cost-raising effect on rivals. One method by which the networks could integrate forward is through a direct broadcast satellite system that bypasses local distribution, sending television programs directly to the home. The DBS strategy might also be of interest to cable networks that face an even worse problem than the networks in having to deal with powerful local outlets.

Direct broadcast satellites. Direct broadcast satellites are no longer a pipe dream, though it remains to be seen whether they will be successful. In February 1990, Hughes, NBC, Cablevision, and Rupert Murdoch announced "Sky Cable," a $1 billion joint venture to bring a new generation of pay television to American homes.[32] Sky Cable was supposed to utilize a high-powered, Ku-band satellite with home antennas the size of a "napkin" and to provide up to 108 channels of programming. Sky Cable was abandoned by its partners, however, on the same day in June 1991 that General Motors and Hubbard Broadcasting announced their own direct satellite venture, to be operational in 1994, based on a technology similar to Sky Cable's.

K-Prime, a DBS joint venture of nine cable multiple system operators and General Electric, was also announced in February 1990. It began to offer service in several test markets in October of that year under the name PrimeStar. K-Prime offers a 10-channel service on an existing GE C-band satellite, with plans to move to a higher power Ku-band service in the mid-1990s. C-band is the existing frequency band used by "low-power" satellites to broadcast to cable systems and home satellite dishes. The K-Prime Ku-band system will use antennas about one meter in diameter.[33] Touchstone Video Network Entertainment Corporation has a national C-band DBS service with pay-per-view movies and six basic services.[34]

Finally, SkyPix announced plans to offer a pay movie service along with rebroadcast of several superstations starting during the summer of 1991. SkyPix plans to use a compression technique to put a number of channels on each transponder of a Ku-band satellite. In theory, digital compression techniques could provide two hundred or more channels per satellite on a DBS system or on a cable television system.[35]

The FCC is not standing in the way of this potential new technology; it seems willing to provide licenses to all who want to give DBS a try. Eight licenses had been awarded as of mid-1991. The substantial capital cost of providing receiving equipment to millions of homes is a more serious problem, as is the problem of finding ways to distribute such hardware through local outlets. Problems of programming multichannel networks such as a DBS system are discussed in Chapter 4. For a useful survey of the problems and potential of direct broadcast satellites, see Johnson and Castleman (1991).

Vertical integration by network rivals. Vertical integration is a strategy followed by the networks' suppliers and competitors. Vertical integration by cable television multiple system operators is discussed in Chapter 6. Extensive vertical integration exists among the networks' competitors (see Table 5.13). Another form of vertical integration is possible between networks and hardware manufacturers. The same logic once led RCA, a radio set manufacturer, to operate NBC radio and later television networks. Other examples of vertical integration include Sony's acquisition of a record company from CBS and a movie studio (Columbia) from Coca-Cola, and Matsushita's takeover of MCA, the owner of another movie studio (Universal).

Tap Viewer Willingness-to-Pay

Broadcast networks produce audiences to sell to advertisers. They might charge viewers in addition to or instead of advertisers, just as cable networks do. This might involve conversion of over-the-air broadcasts to pay broadcasts using scramblers already developed for cable, or perhaps dual broadcast of programs—by cable without ads for pay and over-the-air with ads and without viewer charge, for example. The last published study reported that viewers were

Table 5.13 Vertical integration in the U.S. video industry

Movie studio	Program production	Theatrical films production/ distribution	Broadcast syndication	Cable programming network ownership	Cable systems ownership	Broadcast television ownership	Video cassettes/ disks
Columbia Pictures/Sony	X	X	X				X
Cox	X		X	X	X	X	
Disney Productions	X	X	X	X		X	
Great American Communications	X	X	X			X	
MCA/Universal	X	X	X	X		X	
MGM/UA-Pathe	X	X	X				X
Multimedia	X		X	X	X	X	
Orion	X	X	X				X
Paramount Communications	X	X	X	X		X	X
Time-Warner	X	X	X	X	X		X
Turner Broadcasting	X		X	X		X	
20th Century Fox/ News Corporation	X	X	X			X	X
Viacom	X	X	X	X	X	X	
Westinghouse	X		X	X		X	

willing to pay about seven times more than advertisers for network broadcasts *with commercials* (Noll, Peck, and McGowan, 1973).

Summary

Networks are distributors: they buy a product and resell it. They also assume risks in the process. Networks exist for two purposes: to facilitate sharing the fixed costs of program production and to aggregate local audiences for sale to national advertisers. In other words, not only do networks buy programming for resale to television stations or cable systems, but they also buy audiences from the stations or systems for resale to advertisers. In each case the networks reduce transaction costs by making it unnecessary for hundreds of advertisers and dozens of producers to deal individually with hundreds of television stations. In neither case, however, are networks mere resellers or mere holders of inventory. They add value to both of the products they distribute through packaging and risk bearing.

Despite viewer "grazing" among channels with the use of remote control devices, the audience for a program still depends to a great degree on the popularity of preceding programs on that channel. The packaging of television programs in a network schedule, therefore, involves considerations of audience flow and depends critically on the use of promotional announcements. Similarly, broadcast networks and basic cable networks sell to advertisers packages of commercial minutes or, more specifically, periods of access to audiences of given sizes and demographic composition, typically spread out over the network schedule. In the case of advertisers, the networks bear at least some of the risk that audiences for some programs will be less than expected. In the case of programs, except where federal regulations prevent them from doing so, the networks tend to bear most of the risk of failure, because much of the controllable portion of the risk arises from the scheduling decisions that they make.

Advertisers seek access to audiences of potential buyers of their products. Advertisers' target audiences tend to be specialized. But certain important advertisers of products that are widely purchased seek large, undifferentiated audiences. Until recently, the advantage of the broadcast networks lay in their unique access to the very largest and most undifferentiated audiences. But now competition from new media has fragmented these audiences and therefore reduced the ability of the broadcast networks to deliver this product. As a result, the broadcast networks have more difficulty differentiating their advertising products from those of rivals.

Because of the high fixed costs of programs, large and therefore relatively undifferentiated audiences are cheaper to produce, per unit, than are smaller and more specialized ones. In addition, use of large audiences may avoid wasteful overlap or duplication of reach that arises when more specialized media are used. At the same time, however, large audiences tend to include numerous people who are not potential purchasers of the advertised product. Money spent reaching nonbuyers is wasted. Advertisers must decide whether the benefits warrant buying a big audience: a larger portion of their expenditure is wasted with a big audience, but such an audience has lower unit cost and nonduplication advantages. Competition among media tends to make the effective price (per member of an advertiser's target audience) of small, specialized audiences equal to that of large, undifferentiated audiences. Advertisers who are on the margin between the two create price discipline that benefits those advertisers who are not. One proof of this was the failure of the NAB Code, an agreement nominally restricting the output of ad minutes prior to 1980, to have any actual effect on output.

The reduced ability of the broadcast networks to produce very large audiences means that there is a much smaller advantage for advertisers in buying the largest available audiences: the broadcast networks have higher unit costs and greater audience duplication than used to be the case. Because of this, certain advertisers are actually worse off despite the increased competition; a product they would prefer to buy simply is no longer available in the marketplace.

The networks face strategic problems in their downstream vertical relationships. Affiliated television stations (in the case of broadcast networks) or cable systems (in the case of basic cable networks) are customers for the packages of programs sold by the networks and sources of supply for the audiences the networks buy for packaging and resale to national advertisers. (Premium cable networks and video cassette distributors have a somewhat different problem because they do not sell audiences; they sell to cable systems and to retail outlets, respectively.) But except for the cassette distributors, networks face imperfections in the competition among the local outlets. Local television outlets are scarce, and there typically is only one local cable system.

FCC regulations prevent broadcast networks from entering into efficient affiliation agreements with stations. Remarkably, in light of these regulations and the incentives on the part of individual stations to offer non-network programming at a higher than optimal level, clearances of network programming in prime time remained high throughout the 1980s. There is greater weakness, however, in the daytime.

To retain a substantial share of the economic gains from the bargains it strikes with local outlets, a network must have alternative ways to reach viewers. Because of new technology, cable and broadcast networks now have increasingly viable ways to bypass any local bottlenecks. These include an increased number of local television outlets with declining UHF handicaps; cable systems, perhaps taking advantage of access channels; and, on the horizon, direct broadcast satellites. Multipoint multichannel distribution service is an alternative in some areas, although only for cable networks. These developments are a mixed blessing for the broadcast networks: they reduce the bargaining power of affiliated stations, but they facilitate entry by competing networks.

At the other end of the vertical chain are the program suppliers. Here the networks face two strategic problems. First, competition among the networks, and between the networks and other entertainment media, tends to bid up the prices and economic rents earned by scarce talent. For those forms of programming where talent cost escalation is greatest, the networks try to find substitutes. They have turned to sports, which may offer reduced risks (that is, more predictable ratings) and news division products, which can be purchased more cheaply on the margin than can entertainment series.[36]

The economic structure of national television news organizations illustrates the fundamental characteristics of the network business. The news-gathering and processing organization must be sizable to produce even one product (one newscast series, for example) at competitive quality levels. Having established such an organization, however, additional products can be produced at low incremental cost. CNN has exploited the economic implications of this by making use of its output not only 24 hours a day but, in different formats, over two cable channels and through the local news broadcasts of numerous television stations. The resulting low unit costs permit CNN to succeed despite relatively small audiences with modest demographic merits.

The second supplier problem faced by broadcast networks, though not yet by cable networks, is the success of the Hollywood studios in dealing with government officials. The networks have been barred from competing with the studios in two important businesses: financing prime-time series programs and supplying distribution services in the syndication market. Because of inefficient risk allocation, the first of these prohibitions raises the cost to the networks of prime-time series and places networks at a competitive disadvantage in serving viewers and advertisers.

Risk management is significant in the network business because only a few of the ideas for entertainment series are sufficiently successful to earn a profit. Approximately 20 percent of the cost to the producer of a prime-time series is based on the expected discounted value of "back-end rights": anticipated profits from the sale of the program to foreign markets and to local stations as reruns. Such profits can be enormous, but they are realized by only a handful of the new shows each year. Nevertheless, in hopes of earning such profits, producers choose to invest in their programs over and above what the broadcast networks pay for an initial network run. The additional investment is spent on higher-quality inputs, which improve the attractiveness of the program and thus its chances of success. The producer must come up with a way to finance the portion of the production cost that the broadcast network does not cover. At one time, the broadcast networks provided this financing in return for a share of the back-end profits. When the FCC made this illegal in 1970, program producers were forced to turn to the Hollywood studios to obtain the necessary funds.

A government policy favoring the studios' retention of the business of financing series production (and of off-network series syndication) has continued for two decades despite the absence of any serious argument that the policy benefits viewers. The effectiveness of studio lobbying in this regard stands as a singular example of the importance of treating government policy as an integral part of the process of gaining industrial competitive advantage.

The broadcast networks in the 1990s will have to complete the painful transition to program forms that can deliver smaller audiences cost effectively. In addition, it will become increasingly important for them to differentiate their advertising market products from those of their cable network rivals. Before long, the broadcast networks may have to charge viewers directly. This difficult task may involve bypassing the local affiliates upon which the networks were originally constructed.

To succeed, business strategies adopted by the networks in the 1990s must acknowledge the broadcast networks' loss of supremacy in the electronic mass media. Networks can no longer afford to shoulder any lingering burdens of the old public service myth, whether imposed by themselves or by the government, if they are to survive as businesses in the decade ahead.

6 / Cable Television

Written with Robert W. Crandall

Cable television is a retail system that distributes television signals from a central office ("head end") along coaxial wires or fiber optic cables to individual homes. Local cable systems carry television signals from a variety of sources: national cable networks and both local and distant broadcast stations. To charge customers for special channels, an electronic device in each home is programmed to block out signals for which no payment is made. Newer cable systems can provide certain interactive services. (For example, viewers' opinions can be polled.) But unlike the telephone system, cable is not a fully switched two-way service.

Cable television was invented around 1948, as a way to improve television reception in rural areas, and during the 1980s it became prevalent in urban areas as well. By 1990 cable was available to about 90 percent of all television households in the United States, and more than 50 percent were subscribing (see Table 6.1 and Figure 6.1). Probably the most important factor in this phenomenal growth is the FCC restriction on the number of over-the-air television channels. No such limits apply to cable. The demand for television by advertisers and consumers far exceeds the number of over-the-air channels allocated by the FCC. A second factor is the proliferation since the early 1970s of domestic communication satellites. By 1990 about 70 cable networks were providing service by satellite at a cost in program expense of $2 billion.

Satellites make it easy and inexpensive for thousands of cable systems around the country to obtain program services such as Home Box Office or Cable News Network. Low-power C-band domestic communication satellites were intended for the distribution of television signals to cable systems, but today more than 2 million backyard antennas are aimed at them—so many that cable

Table 6.1 Cable subscription growth, 1955–1990

Year	Cable subscribers (thousands)
1955	150
1960	650
1965	1,275
1970	4,500
1975	9,800
1980	16,000
1985	32,000
1990	51,000

Source: Television and Cable Factbook: Cable and Services (Washington, D.C.: Warren Publishing, 1990), p. C-384.

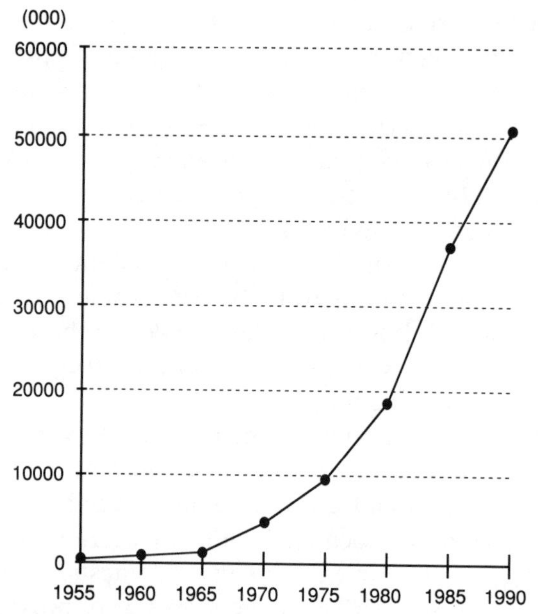

Figure 6.1 Cable subscriber growth

and broadcast networks scramble their signals and attempt to charge homeowners of satellite antennas ("dishes") for descrambling codes. In 1988 Congress passed a law protecting the right of dish owners who live outside the range of ordinary television stations to receive the signals of the broadcast networks. In spite of some piracy problems with cable, it is easier to charge consumers for the use of cable television than for over-the-air television. Viewers value programming much more than advertisers value viewers. Therefore, potential revenue from consumers is more than adequate to cover the substantial capital cost of installing a cable television system. In addition, programming of broadcast network quality can be supported by payments from far fewer viewers than watch ABC, CBS, and NBC.

As discussed in Chapter 5, a number of direct broadcast satellite systems were announced early in 1990, and one is now operational. Cable will face a potentially formidable competitive challenge from this technology, which is not dependent on local retransmission to reach every home in America. For an assessment of the competitive potential of direct broadcast satellites for the cable industry, see Johnson and Castleman (1991).

The new direct broadcast satellites are expected to be less expensive for consumers than the present C-band satellites. But even in 1990 purchasing a backyard satellite dish in order to subscribe to encrypted cable networks was not enormously more expensive than purchasing cable service. Basic cable service in 1990 averaged about $15 per month, or about $180 per year, or about $1,800 in discounted present value at a 10 percent discount rate. The cost of a backyard antenna and associated electronics was between $2,000 and $5,000. The proposed DBS systems would bring the cost of the dish below $500. A satellite dish is able to pick up network signals, but not local broadcasts, so an ordinary television antenna would also be required if viewers had a continuing demand for television programs with local content, such as news.

This chapter focuses on the regulatory and policy issues facing the cable television industry because, like broadcast television in its early years, the shape of the cable industry is largely the result of government policy. Vigorous debate about the proper role of government in the cable industry is likely to continue throughout the 1990s.

Cable and the FCC

Regulatory agencies, faced with a technological or economic threat to their client industries, tend to sacrifice consumer interests to business interests. This pattern has been observed many times in American regulatory history, not only in communications, but also in transportation, securities, insurance, and banking regulation (Owen and Braeutigam, 1978; Noll and Owen, 1983; Peltzman, 1989).[1] Throughout the 1950s, when cable posed little threat to broadcast interests, the FCC denied that it had any regulatory jurisdiction over it. But by the late 1960s, the growing threat to television industry profits had prompted FCC involvement.

Distant-Signal Importation

Toward the end of the 1950s, cable systems began to realize that the demand for their service could be increased if they supplied additional television signals. To obtain these signals they "imported" stations from distant cities. Cable operators chose distant-signal importation to increase demand because it was the cheapest way to acquire additional programming, which attracts subscribers in areas where over-the-air television reception is already good. In the *Fortnightly*[2] case, a 1960s Supreme Court interpretation of the then-prevailing 1909 copyright act, cable operators were not required to pay royalties to the stations originating the signals. Importation was thus cheaper than directly purchasing program materials.

Distant-signal importation affected many parts of the market. In each broadcast market, competition for viewers increased. Local broadcast stations had to compete for viewers who faced many choices. Because an increased number of viewing choices "fragments" the local audience, decreasing advertising revenues for local television stations, the rapid growth of cable threatened to reduce the profitability of virtually every network-affiliated VHF-television station in the nation. Independent VHF stations in large cities, the stations that tended to get transported to other areas by cable operators, had much to gain from the changes. A few became "superstations," imported by satellite to cable systems across the nation and thus able to sell national advertising. The effect of cable on most smaller local UHF independent stations, which broadcast on channels (14 and above) that were difficult to receive clearly on televisions of the 1950s and 1960s, was to improve reception and reduce the difficulty faced by viewers in tuning in the stations. Consequently, UHF audiences increased. But VHF network affiliates had little to gain from cable, and it was this group (allied with the networks) that was the most powerful politically and economically. Accordingly, broadcasters in the 1960s began a successful campaign of active political opposition to cable and, especially, to distant-signal importation.

The response of the FCC was, first, to assert regulatory jurisdiction over cable and, second, to freeze the growth of cable from 1968 to 1972 by banning distant-signal importation in the hundred largest television markets.[3] The first step was justified by the doctrine that the regulation of cable was "ancillary" to television regulation. The second move was protectionist, prompted by an imagined threat by cable to the FCC's rather half-hearted policy to promote the expansion of UHF television stations. But because it improved their reception, cable helped rather than hurt UHF stations (Park, 1972), except in those areas (the smaller markets) where the FCC had not forbidden signal importation.

The Regulation-Deregulation Shuffle

During the freeze from 1968 and 1972, cable systems began to organize their own political lobby, and eventually a compromise was reached that allowed some importation of distant signals after 1972. In 1972 the FCC asserted full regulatory jurisdiction over cable. Cable systems were given various public service obligations, including the donation of free channels, termed "public access" channels, to civic purposes. This "taxation by regulation" (Posner, 1971) has frequently been the quid pro quo for regulatory protectionism, but cable was not given any regulatory "protection" by the FCC; it was merely allowed freedom to compete with broadcasters.

The FCC's 1972 cable decision was a compromise brokered by the Nixon White House among three industrial interests: the major broadcasters, the cable system owners, and the Hollywood studios. Except to the extent that the authors of the compromise sought to end the complete freeze on cable growth, consumer interests were not represented at the table. Issues concerning cable program content, pay cable television, and cable's copyright liability were left unresolved.

The FCC had restricted pay television and pay cable television in 1968, and these restrictions stood until 1976, when they were struck down by the courts. Also in that year Congress passed a comprehensive new copyright act, awarding cable television systems a compulsory license to retransmit broadcast signals in return for a fixed percentage of cable revenues. This money was put in a fund administered by a new federal agency, the Copyright Royalty Tribunal. The act gave cable systems the right to carry all local broadcast signals without payment to the owners of the programs being broadcast. The compulsory free license to carry local signals remains in place in the early 1990s, although it is the subject of growing controversy. In 1979 and 1980 the FCC abolished all of its restrictive rules regarding cable television except the "must carry" rule, which required cable systems to carry all local television signals.[4] The 1984 Cable Communications Policy Act further deregulated cable, removing even the ability of municipalities to regulate rates.[5] In 1987 the courts declared the must carry rule to be unconstitutional, and the last vestige of the protectionist federal regulation disappeared.

A reaction followed. Deregulation of rates led to consumer complaints about the price and the quality of cable service. The broadcast industry began to question whether cable operators should continue to enjoy a free compulsory license to use local broadcast signals. Congressional hearings took place, with threats of anticable legislation. Among other problems, policymakers have become concerned about large cable operators' ownership of cable networks. An important debate has arisen within the FCC and Congress regarding whether cable operators have significant long-term market power in the provision of video transmission service.[6] Only if cable systems have substantial market power can any question arise of anticompetitive effects from buying power in program markets, control of program content, or other forms of vertical integration. The presence of local market power is also a prerequisite to the legitimacy of calls for greater regulation of the rates charged by cable systems to their subscribers.

Were the FCC's early efforts to protect broadcasters' profits from cable television competition justified by the Commission's public interest considerations? Cable is a technology of communications that is nothing more than a different means of delivering television signals, and it has the potential of increasing the number of channels of communication. Having more channels available can hardly make viewers worse off. We know that the supply of programming to fill the new channels is elastic (responsive to price).

The FCC policies did not protect viewers from any serious risk of harm. One group of corporations simply earned, for a time, profits at the expense of another group. The real losers were members of the viewing public in whose name, ironically, the FCC advanced its policies.

New Program Services

The cable industry faced new problems as its regulatory constraints faded away in the 1980s. Rural communities with poor over-the-air reception were wired early. Most urban areas already enjoyed good television reception. If cable was to succeed in these areas, it had to offer new and different services to entice subscribers. Distant broadcast signals did not provide a long-lasting answer, although for a time superstations, such as Ted Turner's WTBS in Atlanta, were an important staple of cable offerings. One source of programming was provided by premium services, for which subscribers usually pay an extra monthly fee. The first premium service was HBO, which specialized in movies that had not yet been shown on broadcast television.

Home Box Office, a unit of Time Inc., pioneered the idea of opening a "window" for cable exhibition in the distribution of motion pictures. Expensive, high-quality motion pictures were "released" to cable television after they had been in theaters but before they were made available to the television networks. Fearing concentration of buying power in the market for cable releases of feature films and other entertainment products, the movie studios formed a joint venture called "Premiere" to distribute films to cable systems. The studios' fear was based on their experience in selling to the three broadcast networks. Premiere did not succeed because of antitrust objections by the Justice Department, which argued successfully that the studios' collective attempt to enter the cable network business could reduce competition in the production of movies. The studios' strategic fears were misplaced in any event. Unlike the broadcast networks before regulatory reform, the cable networks were not protected by FCC policies from competitive entry. There was little danger that the cable network business would be highly concentrated. Although individual niches (movies, sports, music, travel) may have relatively few occupants at any one time, the unconcentrated market in satellite transponders and in the other resources needed by new entrants means that there are not any substantial barriers to entry into these niches, and there is competition for viewers and program resources across the niches.[7]

With cable's success in wiring urban areas came a proliferation of new program services. Many imitated the format and offerings of the broadcast networks. Others specialized, with the most successful early specializations occurring in news (CNN) and sports (ESPN).

The simultaneous growth of cable systems, which depended on new cable networks for the programs that would attract urban subscribers, and of the cable networks themselves illustrates how complex infrastructures of industries with interrelated demands for complementary products and services can arise without central planning, government standard setting, or even vertical integration. Such integration developed for the most part after cable got off the ground. The most fundamental form of vertical integration in the cable television business occurs through the nearly universal practice of bundling.

Bundling of Transmission and Content

Local cable operators—or the multiple system operators (MSOs) that control them—decide which program services to purchase. They bundle these services together with local transmission service and resell this "package" to consumers. A transmission service could be offered on an unbundled basis. For example, a cable operator could set a price at which it would retransmit the HBO movie channel over its system to those subscribers electing to pay for it and collect fees from subscribers on behalf of HBO. Since the break-up of the Bell System, local telephone companies have performed this local transmission and billing function for competing long-distance telephone services. To determine whether unbundling local cable transmission from programming is advisable, cable operators and policymakers must evaluate both the costs and the benefits of bundling the medium together with the message. We begin with the economic reasons for bundling in the cable business. (For a more general review of the economics of bundling, see Winston et al., 1990; for a discussion of programming multichannel bundles, see Chapter 4).

First, if transmission service is sold separately from content, and if individual channels are sold in separate units (but not necessarily as single channels), the cost of service will increase. Most cable operators would have to make a nontrivial capital investment if they were to offer individual channels, or groups of channels, on an unbundled basis to subscribers. There would also be additional costs of billing customers for content on behalf of cable networks. These transaction costs might be large relative to the prices that consumers would be asked to pay for each channel. But, as the common practice of charging extra for premium channels or tiers demonstrates, unbundling is certainly feasible. (If compression techniques become widespread, so that cable systems offer hundreds of channels, some form of unbundling will almost certainly be a necessity.[8] It seems most likely that many programs will be purchased on an individual basis—pay-per-view, or in relatively small groups or tiers of channels.)

Second, there might be marketing problems in unbundling content from transmission or channels from each other. The availability of a particular program enhances the demand for the transmission service. For example, if local newspaper ads promoting HBO attract new cable subscribers, the audience for another cable channel such as MTV may increase. These spill-over effects mean that local promotion of an unbundled channel by its owner might be insufficient to maximize the value of the system as a whole. In addition, there might be cost savings from coordinating the joint promotion of individual channels and transmission service.

Spill-over effects provide a third reason for bundling content with transmission and channels with each other. In an unbundled system, a change in the price charged to subscribers for a given program service will affect not merely the demand for that service but also the demand for transmission, and possibly the demand for complementary program services. In some circumstances strong complementarities between services might lead to less efficient pricing if the services were sold independently.

A fourth reason for bundling content with transmission is that bundling (or other marketing packages) may enable the cable operator to get more money from the subscriber than would be possible if each service had to be priced separately. Bundling offers an opportunity to create packages that result in a kind of price discrimination among consumers. This opportunity obviously benefits the cable operator; the overall effect on economic efficiency is ambiguous (Carlton and Perloff, 1990, chap. 15).

Finally, control of content by the cable operator may help the operator avoid duplication of channels that serve particular consumer preferences. If 50 percent of the people want to watch a news channel, and 10 percent want a health channel, the cable operator may maximize penetration by offering each, while two or three independent advertiser-supported cable networks may each produce news. An independent competitive market for cable network services may not produce a health channel until the news channel has been duplicated many times over. With limited channels, particularly early in the development of its system, a cable operator may need to control content in order to ensure the availability of such a service.

None of the foregoing reasons for bundling necessarily impairs competition or economic efficiency, and some promote it. But there are other, potentially anticompetitive, gains that might accrue to cable operators from this kind of vertical integration at the local level. For example, suppose a local cable system with substantial but not complete monopoly power acquires exclusive local rights to a number of cable networks. Is this anticompetitive? From the point of view of a cable network, the decision to sell rights on an exclusive basis requires a calculation of the lost revenues from sales to multiple local distributors. In addition, there are free rider problems that arise when multiple local distributors each engage in promoting the product. If the supply of network program services is as elastic as it appears, it is unlikely that it would be profitable over the long run for cable systems to attempt to buy up all the available rights in order to deny them to competitors. Moreover, because exclusivity may actually enhance the supply of program services, it would be difficult to justify a ban on such arrangements.

There is also the problem of deals by cable operators to exclude or discriminate against rivals of particular program services. Because video programs are public goods—that is, they have substantial fixed costs but very low marginal costs—a cable network such as Lifetime has almost the same costs whether it has 10 million subscribers or 20 million subscribers. When cable networks compete, either in selling advertising or in selling directly to consumers, the network with the most subscribers gains a great advantage. The more subscribers there are, the greater will be the network's advertising revenue. Services with access to more potential subscribers are in a position, other things being equal, to make profits at prices their rivals may find unprofitable.

The cable network that induces some cable operators to exclude rival networks imposes on those networks the necessity to charge a higher price merely to break even. This gives the network that induces the exclusion a competitive advantage in selling its services to subscribers on systems where the exclusion has not taken place. Exclusion provides a benefit to the cable network in the form of higher profits; it is difficult to find any benefits to consumers in such activity. (A similar analysis applies to discrimination in the pricing and promotion of rival networks.) The cost to the cable network of obtaining higher profits in this way is the amount it must pay to those cable operators who are induced to exclude its rivals.

The cable operator that agrees to exclude rivals of the cable network in question, or to discriminate, incurs a cost penalty, which sets a floor on what the cable network must pay to induce such cooperation. The penalty is the lost subscriber revenue as a result of either the reduction in the quantity of programming that is offered or the distortion of pricing or promotion decisions with respect to other networks. A system offering fewer cable networks or less than competitive levels of promotion will, in general, have fewer subscribers or lower prices or both.

In sum, a network can in theory profitably enter into an agreement with some cable operators to exclude a category of competing networks, such as movie channels, provided there are barriers to entry into the networking business.[9] If there are no substantial barriers to entry or exit, it is pointless to spend resources to exclude one competitor because it will simply be replaced by another. Empirical evidence, in fact, shows that such exclusionary behavior does not occur. Three useful points can be made, however, even absent empirical evidence.

First, although this issue often is thought of as arising in the context of vertical integration of cable operators with cable network program services, it is not a problem unique to vertical integration. If it would be profitable for a vertically integrated cable operator to exclude rivals of a network that it owned, then it would probably be profitable for an independent owner of that network to pay cable systems to be their exclusive distributors. Given the possible economies of vertical integration described earlier, it does not seem sensible to ban vertical integration; exclusion could take place anyway through other means.

Second, the supply of programming to create and expand cable networks is extremely elastic. In the early 1990s more than 70 national cable networks exist. The factors of production used to create these networks consist of talent and communication hardware, both of which are available for rent in highly organized markets where these factors have many other uses. The supply of such services to the cable industry is likely to be very elastic. In these circumstances, it is doubtful that a particular cable network could gain, except possibly in the short run, from excluding or raising the break-even prices of a few targeted rivals. New networks would enter, or existing networks would reposition themselves, to become closer substitutes for the network instigating the exclusion.

Third, in circumstances of easy entry such as those prevailing in the cable industry, individual networks that are in the initial phases of attempting to gather a significant base of subscribers may very well seek exclusive arrangements with cable operators in order to enhance their chances of commercial success. Whether these exclusive arrangements are likely to be commercially profitable, given the prices that will have to be paid to cable operators to induce them to enter into the necessary arrangements, is an open question. But whatever their commercial viability, such agreements cannot possibly pose a significant threat to competition. These observations on the likelihood of exclusionary behavior by vertically integrated multiple system cable operators lead us to the empirical evidence presented in the next two sections.

Natural Monopoly and Local Market Power

Cable television is now available to more than 90 percent of all television households, and more than half of all television households subscribe. In most areas, cable service is provided by a single firm. Because many cable subscribers no longer have rooftop antennas, they do not receive regular over-the-air television broadcasts. Several courts have held, or assumed, that cable television is a natural monopoly (Brenner, 1988a).

A natural monopoly is a market in which there is room for only one firm of efficient size. Natural monopolies are only interesting if the market they occupy is one in which consumers or suppliers have no reasonable alternative to them. In other words, the fact that only one gas station can occupy a given street corner is not a good example of natural monopoly because consumers can readily go to gas stations on nearby street corners. Moreover, not all natural monopolies have "market power," which is the ability over an extended period to charge prices above competitive levels. In some circumstances, even though there is room for only one firm in a market, firms may compete for the privilege of being that firm. This competition "for the market" may drive prices down to competitive levels despite the fact that there is only one actual seller. Another way to describe this process is to say that in some markets "potential" competitors are always ready to move in and to replace the existing seller; as a result, the incumbent dares not charge a price that returns an above-normal profit.

The preceding discussion assumes that competition "for the market" takes place in the form of offering better prices or better service to consumers. A different form, however, is possible: competition for the right to hold a monopoly franchise from the government. In this case, competition may involve bribes, political favors, and expenditures on services such as television coverage of city council meetings for which there is little consumer demand. This form of competition is termed "rent seeking"; it is generally unproductive and wasteful (Bhagwati, 1982).

Is cable television a natural monopoly? The first issue, on the demand side, is whether customers and suppliers have reasonable alternatives to cable service. Most viewers have the alternative of using "rabbit ears" or a rooftop antenna to receive local over-the-air signals carried on the cable. To receive the nonbroadcast channels, such as HBO, viewers can purchase a rather expensive backyard or rooftop satellite dish and pay monthly charges to descramble these signals. The monthly charges are about the same as those paid for premium cable channels by cable subscribers, although some claim that the prices charged to dish owners are too high. (The two leading firms supplying satellite subscription services to dish owners, Viacom and TCI, also own cable systems.) All viewers may also rent prerecorded video cassettes, which generally duplicate the movies available on the premium cable networks, but not the news or sports programs available on the other cable networks. In a few areas there are over-the-air services—wireless cable or multipoint multichannel distribution systems (MMDS)—that can supply a small subset of the cable networks to special rooftop antennas. Finally, there are direct broadcast satellites available, attempting to offer services—especially movies and superstation broadcasts—that substitute for cable programming.

The alternatives of suppliers mirror those of viewers. Broadcasters seeking to reach local audiences can do so over the air, provided cable subscribers have not removed their antennas. National cable networks can reach viewers through direct satellite broadcasts to backyard dishes or, in the future, to smaller receivers via Ku-band DBS systems, or through MMDS in those areas where MMDS is available. Movie distributors can reach consumers through retail video outlets.

In some instances local real estate interests (subdivision developers and apartment complex owners) compete to supply cable service as part of the amenities associated with their residential housing packages. These cable systems, sometimes called satellite master antenna television or SMATV, offer some competition for franchised cable systems, particularly in growing areas.

Are these alternatives sufficiently good substitutes for cable service that a cable owner would not be able to increase prices significantly above the level that returns a "normal" profit? This is an empirical question, the answer to which probably has varied over time and from one city to another. As cable penetration grows and more subscribers abandon rooftop antennas and come to depend on cable networks, over-the-air service becomes a decreasingly attractive substitute for cable service. Finally, as noted earlier, in 1984 Congress enacted a law deregulating cable television in cities where the FCC determined that there was "effective competition" for cable.

The Effective Competition Standard

Since 1985 the FCC's standard for effective competition has been the existence of at least three over-the-air broadcast signals. In 1990 the Commission reviewed this standard and considered proposals to increase the number of local signals required.[10] Generally, the Commission is expected to increase (probably to six) the number of local over-the-air signals that must be present in order for cable systems to escape the jurisdiction of local rate regulators.

GAO Study

A good deal of the debate on the FCC's standard for effective competition seemed to turn on whether cable television's market power has increased since 1984, when Congress passed the Cable Act. Evidence on this point prior to an FCC (1990) inquiry was limited to a survey by the U.S. General Accounting Office (1989).[11] The GAO found that prices for basic services increased between 1986 and 1988. That price increase evaporates when corrected for inflation and for increases in the number and quality of basic services offered. The GAO survey reported that average monthly rates for basic service increased 26 percent, from \$11.70 to \$14.77, between December 1986 and October 1988. The average number of channels offered also increased over this period, from 27 to about 32. Adjusting for inflation and for the number of channels offered, the real price of basic service per channel (in constant 1982 dollars) has remained virtually unchanged at about \$.38.

The number of channels is not the only indicator of the increased quality of service offered to cable television subscribers. Over the same period that nominal prices increased 26 percent, programming expenses of basic cable networks increased 45 percent, from \$513.1 million to \$744.9 million.[12] The GAO study also reported that the nominal price of premium services decreased slightly (and so with inflation, the real price decreased even more) over the 1986–1988 period.[13] Programming expenses for premium services increased 17 percent, from \$848 million to \$989 million, over the same period.[14] Program services are public goods, and an increase in expenditure generally translates into an increase in program quality, from a viewer's perspective, because more popular programs cost more to supply. Thus if any allowance is made for quality improvements, the real unit price of cable service has declined.

The q-Ratio

In the debate over the FCC's standard of effective competition, much was made of the "q-ratio" measure of cable market power. The q-ratio is equal to the market value of an asset divided by its reproduction cost. The notion is that in a market in competitive equilibrium, with no monopoly rents, the market value of a firm's assets should tend to equal their reproduction cost. But in fact a high q-ratio does not necessarily mean that a firm is making high profits, and high profits do not necessarily mean that a firm has market power. Although a useful measure of profitability, the q-ratio is "subject to large individual errors and may often be seriously misleading" (McFarland, 1988, p. 622).

Even if a high q-ratio indicates high profits, those profits may not reflect entrenched market power. In the U.S. economy, high profits serve as a reward for economically desirable behavior. In particular, high profits reward superior efficiency and competitive risk taking. Thus a firm's high q-ratio could merely indicate that it is particularly efficient. Alternatively, a firm's high q-ratio could indicate that its industry is in disequilibrium and needs additional investment to satisfy consumer demand (Shrieves, 1987). As investment in the industry grows, profits and q-ratios will fall, but this investment may take many years. The fact that high q-ratios may merely signal a need for additional investment is a particularly important consideration in a new and fast-growing industry.

Increases in Output

A case could be made for the reimposition of regulation if real unit prices for cable service had increased while output (customarily though very inadequately measured by number of subscribers or by number of channels of programming) had decreased. But output has dramatically increased since deregulation. Between 1985 and 1990, the number of nationwide subscribers to basic cable service increased 38 percent—from 38 million subscribers to 52.3 million subscribers; basic cable services distributed by satellite nationwide increased 65 percent during this five-year period—from 34 services to 56 services.[15] There is thus no evidence that the FCC's effective competition standard, which led to deregulation of basic cable rates, made consumers worse off. Moreover, there can be no question that in the 1990s cable operators face more competition from other distribution technologies. Since deregulation, there are many more SMATV and MMDS subscribers. The number of VCR households and home satellite dish (HSD) owners has also increased, and DBS services have now been launched.[16]

No Case for Reregulation

Changes that would warrant reregulating the cable industry are lacking. The FCC's conclusion in 1985 that three or more over-the-air signals constrain cable operators' market power in providing basic service was based on a staff study of viewing patterns involving nonbroadcast programming.[17] The FCC's theory was that as long as most cable viewing involved broadcast signals, local over-the-air broadcast service was likely to be an effective substitute for cable service. (Of course nonbroadcast programming on other media, such as video cassettes, and broadcast programming available on SMATV and MMDS systems could be strong substitutes as well.) The FCC found that on cable systems competing with three over-the-air broadcast stations, the viewing share of such nonbroadcast programming ranged from 23.5 percent to 34.5 percent, based on 1982 data.[18] More recently, the FCC reexamined the same cable systems using 1988 data (FCC, 1990). The viewing share of nonbroadcast programming ranged from 19 percent to 36 percent—remarkably little change from the previous data.

It is a matter of interpretation whether the viewing share had increased or decreased; in either case, given the size of the sample (11 systems), it is very doubtful that any statistically significant change occurred. Viewing shares for systems with two, four, and five competing broadcast signals were also very similar in 1982 and in 1988.

In its 1990 notice of proposed revisions to the effective competition standard, the FCC asserted that retaining a standard based on broadcast signals "presumes that today's basic cable subscriber chooses basic cable service primarily to get better reception of local television signals rather than to receive cable programming not otherwise locally available on broadcast television."[19] In reality the relevance of broadcast signals as a source of cable competition rests on no such presumption. Although cable systems have increased the amount and quality of nonbroadcast programming offered as part of basic service, cable subscribers still spend the bulk of their viewing time watching programming that is also carried by broadcast stations. According to recent estimates, the viewing share in all cable households of network-affiliated, independent, and public television stations is 66 percent.[20]

In research sponsored by the National Cable Television Association, Dertouzos and Wildman (1990) found that cable systems facing competition from five or more broadcast signals had fewer subscribers and carried more basic cable networks than did systems facing competition from fewer than five broadcast signals. They also found that the price of basic service per basic channel was significantly lower for systems competing with five or more broadcast signals. Increasing the number of competing broadcast signals beyond five did not have any significant additional effect on cable systems' performance, however.

In a study sponsored by TCI, the largest multisystem owner of cable, Crandall (1990c) confirms that the performance of cable systems is affected by competition from broadcast signals. His results were similar to those obtained by Dertouzos and Wildman despite differences in the source and size of the data base. Dertouzos and Wildman (1990, pp. 9–11) used a sample of 340 cable systems and drew their data principally from A. C. Nielsen. Crandall's study relied primarily on a sample of 2,752 cable systems providing 1989 data to the *Television and Cable Factbook.*

A more significant difference between the two studies lies in their measurement of competition from over-the-air signals. Dertouzos and Wildman (1990, p. 13) measured the number of locally available over-the-air signals by counting the number of television stations that included within their Grade B contour the area served by each cable system. This measure is especially useful because it corresponds to the measure the FCC has used to determine whether effective competition is present. Because of the large size of the sample in Crandall's study, it was not feasible to count the number of Grade B signal contours surrounding each cable service area. As a proxy measure, Crandall used the number of broadcast channels that each cable system receives over the air. A comparison of the two measures developed for a portion of the data showed that the two measures are highly correlated.[21]

Crandall controlled for the effects on price of various system characteristics (such as channel capacity, miles of cable, homes passed, number of cable neworks carried, and the number of subscribers) through regression analysis. Demographic and economic factors, such as income levels and the age distribution of the population, were also included in the regression analysis. Controlling for all these factors, the regression estimates how the price of basic service for one group of systems differs from the price of basic service for another group of systems.

Crandall's results are presented in Table 6.2. He found, for example, that cable systems with two or more over-the-air channels charge $1.45 less for basic service than do systems with only one over-the-air channel, holding other factors constant. This estimated difference is statistically very significant.[22] There are two striking features of the price effects in Table 6.2. First, the number of over-the-air channels appears to make a significant difference in the price of basic cable service, even at a low number of channels. Second, the price differences between the test group and the comparison group appear to decline and then become insignificant as systems with additional signals are added to the test group. The declining pattern of price differences implies that the incremental effect on price of an additional local signal declines as the number of broadcast signals increases. When the test group is modified to include cable systems with six or more over-the-air channels, the price difference is no longer significant, and it remains insignificant as systems with still more over-the-air channels are added. This implies that there is some threshold number of broadcast signals that provides substantially all of the competitive effect on the basic fee.

Table 6.2 Effect of broadcast signals on price of basic cable service

Over-the-air channels			Effect on basic service price in test group ($ per mo.)	*t*-statistic
Test group		Comparison group		
2 or more	vs.	1	−1.45	−2.96*
3 or more	vs.	2 or fewer	−0.85	−2.66*
4 or more	vs.	3 or fewer	−0.82	−3.73*
5 or more	vs.	4 or fewer	−0.46	−3.00*
6 or more	vs.	5 or fewer	−0.15	−1.08
7 or more	vs.	6 or fewer	−0.20	−1.47
8 or more	vs.	7 or fewer	−0.21	−1.54
9 or more	vs.	8 or fewer	−0.29	−1.89

Source: Crandall, 1990c, p. 21.

*The *t*-statistic is significant at a 95 percent confidence level. Note that *t*-statistics are reported for the log of price. Price effects have been evaluated at the sample mean.

These results provide support for two basic features of an effective competition standard based on broadcast signals. First, the number of broadcast signals clearly affects the price of basic service, which is the aspect of cable systems' performance that has received the most attention. Second, there appears to be a threshold beyond which additional broadcast signals have no significant effect on price. These results do not identify exactly the number of Grade B broadcast signals at which this threshold occurs, because in this analysis the number of broadcast signals has been represented using a proxy variable. There is, however, some number of broadcast signals that provides substantially all of the competitive effects that would be derived if more signals were present. Dertouzos and Wildman find that this threshold occurs at the level of five broadcast signals, a result that is consistent with Crandall's analysis.[23]

Price Elasticity of Cable Demand

Another perspective on cable market power stems from measures of demand elasticity. The elasticity of demand for cable service is the percentage decrease in subscribers that would result from a 1 percent increase in price.[24] If the elasticity of demand is between zero and one, demand is said to be inelastic. If elasticity is greater than one, demand is said to be elastic. The more elastic is demand, the less profitable it will be to increase price.

The elasticity of demand is an important factor in determining whether a firm can profitably raise its price. When demand is inelastic, it always pays to raise prices. A price increase will increase total revenue, because the increase in price will more than offset the decrease in subscribers. A firm will continue to raise its price up to the level at which further price increases are not profitable. Where this occurs depends on the firm's cost structure. When demand is elastic, a price increase results in lower total revenue, because the quantity reduction is larger than the price increase. Profits may continue to increase, however, because the decrease in quantity also reduces the firm's operating costs. The profit-maximizing price is achieved when a price increase is just balanced by the decrease in quantity and cost, or when marginal revenue equals marginal cost. Note that the relevant cost is the cost of supplying the incremental or marginal quantity that would change in response to the change in price.

At levels of programming prevailing in the early 1990s, it is unlikely that price increases for basic cable service would prove profitable (Crandall, 1990c). This is what one would expect if cable operators are maximizing profits.

Using regression techniques, Crandall (1990d) estimated the elasticity of demand for basic cable service from data on TCI owned-and-operated systems. He found that demand for basic cable subscriptions is relatively elastic.[25] The exact magnitude of the elasticity depends on the choice of factors that affect the price and the quality of basic cable programming. Most of the estimates range from 1.6 to 3.4 (see Table 6.3). The average of the estimates, 2.2, is also the value around which several of the estimates cluster. If the elasticity is 2.2, a price increase of 1 percent for the basic subscription fee for a TCI system—holding constant other factors, such as the quality of programming—would be expected to cause more than 2 percent of customers to discontinue service. To the extent that monthly fees of other cable operators and the quality of service are roughly comparable to those for TCI, these results can be applied to the cable industry as a whole.

It is necessary to examine both price and cost to determine whether it is profitable to raise price, given a level of demand elasticity. When the elasticity is 2.2, a mathematical calculation estabishes that the profit-maximizing price is 1.8 times the cost of producing the marginal quantity (marginal cost).[26] Elasticities of 1.6 and 3.4 imply profit-maximizing prices that are 2.7 and 1.4 times marginal cost, respectively.

To obtain information regarding the prices and costs for an "average" TCI system, Crandall (1990d) examined 1989 TCI cost data. Marginal cost is approximated by the average level of "variable" costs (those that vary with quantity).[27] Costs that pertained to providing premium service only were excluded. The ratio of revenue per subscriber divided by average variable cost per subscriber was 2.0. Crandall found that the estimated price-cost ratio for a representative TCI system of 2.0 was quite close to the profit-maximizing price-cost ratio of 1.8 implied by his estimate of price elasticity (2.2).

Table 6.3 Parameter estimates of demand equations of basic subscribers for TCI systems (two-stage least squares based on different instrument sets; based on 436 TCI systems in 1989; dependent variable is log(number of basic subscribers))

		Estimated parameters of the demand equation						
Instrument set	Adjusted R^2	Constant	Log(price)	$(\text{Log(price)})^2$	Log(homes passed)	Log(major satellite-based networks)	Log(personal income per household)	Elasticity of demand
A	0.924	2.526	−1.760		0.895	0.765	0.059	−1.760
		*1.788**	*−3.948****		*31.976****	*4.022****	*0.534*	*−3.948****
	0.934	−1.057	0.868	−0.570	0.915	0.690	0.134	−2.329†
		−0.488	*0.656*	*−2.094***	*32.633****	*3.787****	*1.212*	*2.826****
B	0.910	3.470	−2.134		0.915	0.693	0.074	−2.134
		*2.072***	*−3.622****		*27.706****	*3.028****	*0.603*	*−3.622****
	0.912	−3.567	2.743	−1.091	0.913	0.845	0.227	−3.375†
		−1.106	*1.367*	*−2.539***	*27.997****	*3.620****	*1.682**	*−2.789****
C	0.931	2.442	−1.578		0.918	0.583	0.049	−1.578
		*1.864**	*−4.118****		*38.370****	*3.712****	*0.463*	*−4.118****
	0.937	−0.632	0.674	−0.504	0.933	0.551	0.119	−2.151†
		−0.329	*0.596*	*−2.105***	*39.042****	*3.657****	*1.108*	*−3.094****

Source: Crandall, 1990d, p. 27.

t-statistics in italics.

† Evaluated at the mean value of log of price across all observations.

* Significant at the 90 percent confidence level; ** significant at the 95 percent confidence level; *** significant at the 99 percent confidence level.

Instrument sets

In all regressions: Constant, log of starting date, log of number of overair signals, log of miles of cable per household, log of percentage of cable miles underground, log of county population, log of employment per population, log of houses passed by system, log of average county salary, log of personal income per household, log of persons per household in county, percentage of county population under 16, percentage of county population between 16 and 20, percentage of county population between 21 and 65, system in second 50 television market, system in small-city television market, system not in city television market. Additional instruments in A: Regional dummy variables for the Southeast, South Central, East North Central, West North Central, Great Plains, Northern Rocky Mountains, Pacific Northwest, and Southwest. Additional instruments in B: Squared terms for all instruments except dummy variables. Additional instruments in C: Regional dummy variables as in A and squared terms as in B.

Competition from New Entrants

If viewers lack good substitutes for cable service, and if most cities are forbidden under the 1984 act to cap cable rates, the only effective constraint on cable prices is competition. One form of competition is "overbuilding," in which a new cable firm competes by running new cables down the same streets where another cable system is already in place. There are between 25 and 50 cities with such overbuilds. According to a survey conducted by Consumers' Research, subscriber bills average $17.31 in nonoverbuild markets but only $14.13 in overbuild markets.[28]

A so-called overbuilder—a potential entrant that wants to compete head to head with an existing cable system—faces serious difficulties. First, cities are often very reluctant to grant competing franchises. Their reasons vary, but among them are fear of disruptions associated with construction, fear of endangering promised benefits to the city government derived from monopoly profits, and perhaps fear of disturbing a political arrangement favoring the incumbent. Second, the new entrant faces the prospect that, because of economies of scale, its costs will be higher than those of the incumbent as long as it has fewer subscribers. The evidence (now somewhat dated) is that two competing cable systems, each serving half the subscribers in an area, will have per subscriber costs about 14 percent higher than a single system serving the same subscribers (Owen and Greenhalgh, 1986). Third, the imcumbent may react to the overbuilder's entry by lowering its price. Perhaps the best outcome from the perspective of the entrant is to be bought out by the incumbent system before any new construction takes place. But the number of potential entrants is very large, and no cable operator wants to gain a reputation for paying off all challengers. Thus a cable operator that seeks to become a profitable overbuilder must establish a reputation as one who completes overbuild projects and who is not easily bought out.

The incumbent cable operator, not wanting to be exploited by potential or actual entrants, has three basic strategies available. The first is to use the politics of the franchising process to discourage or delay new entrants. The second is to establish a reputation as one who never pays off potential entrants and who fights to the death with those who do enter and overbuild. The third is to charge a price and earn profits sufficiently low to discourage new entrants in the first place.

Although an efficient firm will have lower costs than will two or more competing firms, consumers may be better served by competition that ensures that prices will be aligned with costs and that costs themselves will be minimized. Owen and Greenhalgh (1986) estimated a translog cost function for cable television systems based on 1980 franchise bidding data (see Table 6.4). They compared the average and marginal costs, per subscriber and per channel, for a sample of cable systems in 1980–1982, and found that the average system had a per subscriber unit cost of $39.33 as compared with a marginal cost of $30.01 (per year). Although these data reveal a considerable gap between average and marginal cost, this gap reflects the substantial fixed cost of a cable system rather than the declining marginal costs of larger systems. Experience in similar industries suggests that declining average costs (economies of scale) associated with spreading high fixed costs over additional consumers or units of output is not a reliable guide to the existence of a natural monopoly. To take one example, the firm whose costs are being measured may have *chosen* to adopt an unnecessarily expensive or unnecessarily capital-intensive production technology, perhaps to evade regulatory constraints.

One curious feature of the Reagan era of antitrust law enforcement was the decision by the Department of Justice not to challenge horizontal mergers between competing cable systems in the same community (Manishin, 1987). Potential competition from new entrants, though perhaps not a very effective tool for limiting

Table 6.4 Predicted cost per subscriber per month (based on 1980–1982 translog econometric estimate of cable cost function)

	Cost per subscriber		Cost per channel	
	Average	Marginal	Average	Marginal
Average system (30,481 subscribers)	$39.33	$30.01	70.0¢	27.0¢
Large system (90,000 subscribers)	$34.17	$19.08	61.0¢	37.2¢

Source: Owen and Greenhalgh, 1986, p. 77.

monopoly pricing of cable service, may be the most effective force at work until DBS service becomes widely accepted (Hazlett, 1990b; Smiley, 1990).[29] The most likely potential entrants arguably are those cable systems serving contiguous areas. To permit such systems to merge is to reduce further the competitive constraints on cable pricing and service quality.[30]

Too Few Gatekeepers?

Some believe it is desirable to increase the minimum number of entities whose approval is essential to the success of a new program service. To succeed, a new cable network will need access to some portion (perhaps 60 percent) of all cable subscribers. In that case, if more than 40 percent of subscribers are controlled by the top three MSOs, then three key buyers can effectively veto any new service.

It is difficult to make sense of this concern. First, the number of subscribers needed for the "viability" of new services will obviously vary, depending on costs, competing services of that type, advertising demand, and the prices paid for the service by cable systems. Some services seem to be viable despite their access to only a small fraction of cable subscribers. Recent data on the number of subscribers to 56 national basic services show that almost half the services reached less than 25 percent of total subscribers, and almost a quarter of the services reached less than 10 percent.[31] No regional basic services reach as much as 10 percent of total national subscribers, and many reach less than 1 percent. It is easy to imagine a service that might be viable only if it had access to, say, 98 percent of all subscribers; any system with 2 percent or more of the subscribers would be able to "veto" this service. Thus, even if there is a legitimate policy or First Amendment concern involved in the number of gatekeepers, there does not seem to be a principled basis for drawing any particular line.

Second, and more important, this concern for the number of MSO gatekeepers has no logical basis in competition analysis or even in diversity analysis. From the perspective of competition policy, there would be a concern only if concentration of decisionmaking power were accompanied by an economic incentive to restrict output. But it is not. It is in the economic interest of MSOs to encourage new program services, because new program services enhance the demand for cable service. Like any individual system owner, an MSO will prefer a new service that appeals to unserved or underserved consumer interests to a new service that duplicates a service already offered. The former will often enhance demand for cable service more than the latter. In any event, there appears to be no incentive for an unintegrated MSO to refuse to carry a program service likely to increase consumer demand for cable service.

Third, it is important to remember that cable MSOs are by no means the only gatekeepers. Concern about access to ideas and expression cannot reasonably be limited to the video medium, much less the wired video medium. Even if three cable MSOs could determine which video services would be carried on cable nationwide, and even if the MSOs wanted to restrict the number of video services, or to bias their selections, there is no basis to conclude that this would significantly constrict the flow of ideas and expression. Consumers have numerous alternative channels of communication: books, newspapers, broadcast stations, magazines, theaters, and other media.

The Issue of Access

In contrast to what seem to be misplaced concerns about MSO ownership of program sources, the problem of local control of access by cable operators is both more real and more difficult. The "essential facility doctrine," an antitrust concept, provides one perspective from which to view this issue. The essential facility doctrine requires a monopoly owner who is engaged in related competitive markets to give competitors access to the facility in some circumstances. In cable television, access has been a policy issue for years. The Rostow task force on communications policy explored the issue in 1968. The President's Cabinet Committee on Cable Television in 1974 recommended the establishment of a right of access to cable. The House Subcommittee on Communications made a similar recommendation in 1978. And the ACLU and various scholars have also recommended the establishment of a right of access to cable systems (Nadel, 1983; Brennan, 1990). But neither Congress nor the FCC has acted on these recommendations to the extent of requiring cable operators to be common carriers, with no interest in the programming carried. Under the Cable Communications Policy Act of 1984, large cable systems must provide "commercial access" under certain conditions, but this requirement has not led to significant exercise of access rights (FCC, 1990).

Many of those who have advocated a statutory right of access to cable television have done so for reasons based on freedom of expression rather than economic monopoly (see, for example, Owen, 1975). Advocates of access begin with the assumption—thus far unsupported by convincing evidence—that cable television is, or will become, the only medium through which citizens will have access to various sources of news and information, especially local services. Permitting the owner of such a system to decide who shall speak and who shall not, in circumstances where decentralization of that decision appears costless to society, is seen as a potential threat to free local debate. Even more troubling to us is the specter of federal intervention in regulating the content of cable transmissions or in licensing cable owners. Broadcast licensing and content regulation were premised on notions of scarcity and monopoly that are more applicable to a monopoly cable system than to a competitive broadcast station. The courts, however, have generally upheld the First Amendment rights of media owners—even monopolists—over the rights of those demanding access to those media. One can speculate whether the Supreme Court would uphold the constitutional rights of the owner of a hypothetical new technology that monopolized all local or national channels of communication against the rights created by a statute requiring access on reasonable terms. First Amendment scholars hold a range of views on the issue of cable's significance in such a context; for particularly articulate and useful surveys see Brenner (1988a, 1988b).

Especially interesting is the contrast between the "common carrier" obligation of a telephone company transmitting messages, or a railroad transporting newspaper and magazines, and the First Amendment freedom of a cable operator. The common carrier has no right under the First Amendment to censor the messages or media that it carries; yet the First Amendment is said to protect the right of the cable operator to determine content. There is no reason to distinguish among telephone companies, railroads, and cable systems in this respect, assuming equivalent degrees of market power. Asserting their First Amendment rights, telephone companies may attempt to provide cable services in the future, and to control program content.

The context in which the essential facility issue will arise in cable television is likely to be a private antitrust lawsuit. For example, a local broadcast station will allege that it competes with the cable system in selling local advertising, and that it cannot compete effectively unless its signals are carried by the cable system. The owner of the cable system refuses to do so on reasonable terms. These allegations can be generalized. The complainant is a competitor. The defendant competes with the complainant in one market, while owning assets that the complainant requires in order to compete.

To make sense of an essential facilities access case, the assets to which access is demanded must, almost by definition, constitute a monopoly. If they did not, the person demanding access would not be able to show that they were essential because the facilities would be available from some other seller. The concept of monopoly, to be useful here, must be rather strictly defined. The legal definition, which in some courts may encompass market shares as low as 60 percent, clearly will not do; to admit that, say, 40 percent of the market is controlled by others is to admit the possibility of an alternative to the essential facility. Monopoly in the essential facilities context must mean, first, that the monopolist controls 100 percent of a well-defined antitrust market and, second, that significant barriers to entry make it unlikely that a new seller would attempt to enter despite the monopolist's refusal to serve the plaintiff and others who have a demand for access.

"Having a demand for access" means being willing to pay for access a price that at least equals the monopolist's marginal cost of providing access. If the complainant is not willing to pay a price that at least covers the cost to society of providing access, economic efficiency dictates that access should be denied.

Vertical integration is inherent in the essential facility problem. If the owner of the essential facility is not vertically integrated, then denial of access is not denial of a competitor, because the essential facility itself has no competition. Leaving aside certain contractual arrangements equivalent to vertical integration, there is no reasonable basis to expect that discrimination by a nonintegrated monopolist will eliminate efficient competition among its customers; indeed, the monopolist has the opposite incentive. What gives rise to the essential facility problem is the decision of a cable operator to originate programs or to have an interest in the profits from cable advertising.

Why would the owner of an essential facility wish to deny access to a competing firm? Several possibilities occur. The owner may wish to reap monopoly profits in the market where competition is taking place. But because the price that must be paid for such monopolization is forgone revenues at the bottleneck, essential facility stage, this is likely to be profitable only under two special conditions: when such tactics raise rivals' costs and when they facilitate an evasion of regulatory constraint. In addition, exclusion of rivals from access to an essential facility may be profitable if there are economies of vertical integration that result in lower costs when both stages of production are more completely monopolized. (Of course, in this case the integrated firm can increase its market share "naturally" through lower prices, although this may be a slower process.) Finally, to the extent that managers of essential facilities are protected from competition in their product markets and in the market for corporate control, they are free to exercise their discretion by engaging in exclusionary conduct that serves their private interests rather than the interests of the firms they manage.

Those who favor common carrier status for cable have a straightforward economic argument. The local cable system, if it leases channels in the manner of a common carrier, will be incapable of extending its "natural" monopoly of transmission—assuming it has one—backward into "unnatural" control of program selection or production. Instead, there will be competing suppliers of programming, each using the cable system in the way a magazine or book publisher uses the postal service. The only place there could be any concentration of control of content is in the function that corresponds to networking in television or distribution in the motion picture trade. But there does not seem to be any reason to expect significant concentration of economic power in this function.[32] As noted above, there are more than 70 cable networks, and no one company owns more than three or four of them.

With digital compression techniques, which may permit cable systems to carry hundreds of video signals simultaneously, the magazine industry analogy becomes compelling. With pay television and common carrier access, cable systems would have very much the same relationship to program sources as magazines do to the postal service. The cable will serve as a delivery conduit for diverse programs, some specialized, others of mass appeal. But unlike the experience of broadcasters, there would be only minimal content regulation by the state, and the First Amendment would apply with its original force. Neither the cable operator nor anyone else would have significant monopoly power in the market of ideas, and the danger of government censorship would be minimized.

Monopsony Power

If cable systems are assumed to be monopolists in dealing with consumers of local video transmission service, they also probably have power as buyers of program services for resale to consumers. As monopolists of transmission service, they (absent regulation) overcharge consumers of that service. But as consumers of cable network services, they will want to consider the effect, if any, of their purchasing behavior on the prices they pay for those services. Given the manner in which cable networks are produced and operated, it is difficult to see why a local cable operator would regard its decision to purchase an incremental network as affecting the availability or price of other networks. If no such effect exists, local monopsony power translates into mere bargaining power (Crandall, 1990a).

We distinguish between monopsony power and bargaining power here based on whether there is a reduction in output. Whether the exercise of bargaining power is anticompetitive depends on one's view of the purpose of antitrust law. Bargaining power may be unfair, but it is not inefficient, and if the ultimate consumer is not injured, it is difficult to see a legitimate basis for government intervention to cure imbalances of bargaining power between commercial entities.

There is a national market in which program producers compete to sell their products (television programs, movies, prerecorded video cassettes, sports events, and so on) to national or regional distributors (broadcast networks, cable networks, theatrical distributors, direct satellite broadcasters, and video cassette distributors) for packaging, promotion, and resale to local retail outlets (television stations, cable systems, theaters, and video outlets).[33] At the end of this distribution chain are local markets. Cable program services or networks deal with national suppliers on the input side and with local outlets or groups of local outlets on the output side.

The exact boundaries of these markets can be determined, and assessments of market power made, only through an empirical inquiry. For purposes of competition analysis, a market is something worth monopolizing. To paraphrase the language of the Merger Guidelines of the Justice Department, if a hypothetical monopolist of (say) the business of supplying nonbroadcast video program services to cable operators cannot raise prices profitably, then that business does not constitute an antitrust market. There are two reasons, in principle, why such a hypothetical monopoly might not be able profitably to raise prices. First, customers (cable systems) might be able to substitute other sources of supply (for example, broadcast signals) in sufficient quantity that the monopolist would lose too many sales. Second, the attempt to raise prices might simply attract new entrants. Analysis of these possibilities requires facts concerning the alternatives available to customers and the costs and other conditions of entry. The exercise of defining markets in analyzing competition cannot be conducted a priori, and sound competition policy cannot be determined without a basis in market analysis.

A monopsonist is the sole buyer of a product. In order for a monopsony to be of concern from an economic point of view, an important condition must be met: the monopsonist must be in a situation in which each additional unit of a product that it buys increases the price it must pay for *all* the units it buys. This will be true only if the industry that supplies product to the monopsonist has increasing costs, and in that case the monopsonist faces an upward sloping supply curve.

Taking account of an upward sloping supply curve will lead a monopsonist to purchase smaller quantities of the input than is economically efficient. Thus the objection to monopsony from the standpoint of economic efficiency is not that the monopsonist gets a lower price than competing buyers would have to pay, but that overall output is reduced below competitive levels. This reduction leads to the same sort of economic welfare loss as does monopoly.[34]

In practice, monopsony is rarely an important policy problem because the necessary condition described above is seldom satisfied. Most resources used as inputs are not highly specialized in particular end uses. If those resources have other outlets than the monopsonist (that is, if the long-run supply curve facing the monopsonist is flat), then the monopsonist's purchases will have no effect on price or output.

Sources of programming for video distribution vary greatly: new and preexisting entertainment programming of many types, sports, news, public affairs, and locally originated sources. Each type of programming draws on different input resources (such as baseball players, actors, reporters, and animation specialists). The programming supply curve facing the cable industry may be flat in the long run because each of these resources has alternative occupations in addition to the production of programs for cable television. Indeed, if the curve is flat, a cable system or multiple system operator has no incentive to restrict the quantity of programming purchased. For this reason monopsony power of MSOs is not likely to be a serious concern, at least in the long run.

For the purpose of analysis of monopsony power, however, assume that the national supply of programming for cable is not flat, but rather increasing-cost, and thus potentially subject to the exercise of monopsony power. An increased number of cable networks of given popularity can be called forth only by increasing the prices paid for all cable networks.

There are two parts to the analysis. The first is based on the assumption that local cable systems do have monopsony power and hence that they pay less than competitive prices for program services. In the extreme, systems with monopsony power need only offer a price sufficient to cover the marginal cost of delivering the program, which is very low. In this case we show that the adverse effects on viewers that the exercise of this power could cause are actually offset by MSOs.

The second part of the analysis is based on the contrary assumption—that local cable systems do not have monopsony power in the local distribution of video services. Although it is an empirical question, it seems doubtful that they do.[35] Obviously there are other media that compete with cable in buying video products, especially in the long term. In this case we demonstrate that an aggregation of cable systems, none of which individually has buying power, into an MSO (however large) is unlikely to create an organization with monopsony power.

Turning now to the first part of the analysis, assume that local cable systems do have monopsony power (that is, that they do not face effective competition from other local media or from DBS systems). A representative cable system will take the number of services offered as given, but will pay too little when compared with what it would have to pay if it had to compete with other local media for programs. If the supply of programs is not flat, the system's underpayment will reduce slightly the national supply of programs by lowering the returns to being in that business.

A reduction, albeit slight, in the national supply of programs will injure other cable systems by reducing the number of programs among which the other systems can choose (a spillover effect). But no individual system has any economic incentive to take account of this spillover effect on other systems. Every individual system is in this position. In the aggregate the cumulative effect of the spillovers may be a substantial reduction in the supply of programs. Thus the assumption that there is local monopsony power combined with the assumption that the supply of programs is not flat may lead to a substantial market failure and a public policy concern.

Let us now consider the consequences of MSO ownership of cable systems in this situation. A large MSO would notice that its action in paying too little for programs in hundreds of individual systems was having the effect of reducing the supply of programming, resulting in lower profits. The MSO, precisely because of its recognition of its own buying power, would find it profitable to act to expand the supply of programs. If the reduction in supply caused by the problem of local monopsony power were very substantial, the MSO's decision about its purchases would bring output closer to the efficient level than if no MSOs were permitted. In this case MSOs would have monopsony power, but it would be exercised in a benign way, making consumers better off than they would be if thousands of individual systems exercised buying power.[36]

Turning now to the second part of the analysis, assume that local cable systems do not have local monopsony power. That is, assume that local cable systems face effective long-run competition from alternative local media. Does their aggregation into MSOs create monopsony power where it did not already exist?

It is important to realize that no individual cable system competes with cable systems elsewhere for video programs. The nature of television programs is that they can be sold in one area with absolutely no effect on the supply in other areas. From the perspective of a national seller of program services, cable systems serving two different areas are not substitutes; if anything, they are akin to complements in the provision of delivered program services.

If each local cable system has too small a share of each local market for programs to have monopsony power, then an aggregation of all cable systems would not have any greater power in the corresponding national market. The availability of competing local media (and the distributors that serve them) prevents MSOs in this situation from exercising market power. The assumption that local cable systems do not have long-run monopsony power therefore leads to the conclusion that multiple ownership cannot create such a problem, at least from the point of view of economic efficiency.

It has been claimed that large MSOs pay lower prices for certain program services than do smaller systems. Assuming this is true, does it contradict the argument above? It does not, for several reasons. First, MSOs can have monopsony power only if local cable systems have monopsony power, a fact not evident. If local cable systems do not have monopsony power, then low MSO prices cannot be interpreted as evidence of MSO monopsony power, but must signify something else. Second, there are other explanations of the lower prices that do not depend on monopsony power. Any seller of a service calls first on the largest potential buyers, because the selling costs are lower in finding and negotiating with such buyers. A reduction in transaction costs leads to greater profit, which the seller can afford to share with the buyer. The costs of negotiating the terms of a contract to supply program services are not greatly affected by the number of subscribers. Sellers dealing with MSOs with many subscribers, and the MSOs themselves, will find it worthwhile to negotiate amounts and terms that would not make sense for smaller systems. Not only may transaction costs be lower for deals involving MSOs, but the MSOs may have greater bargaining power with respect to the allocation of the profits arising from carriage of a program service. Savings in transaction costs are to be encouraged because they enhance economic efficiency.

Bargaining power is not the same as market power. Market power results in reduced output. Bargaining power merely shifts profits between seller and buyer. There is no basis for policy concern with bargaining power when it does not reduce output. In sum, there is little basis for concern that the buying power of MSOs significantly lessens competition.

Vertical Integration

Cable operators need an ample supply of programming to attract and retain subscribers. Cable networks need access to cable subscribers in order to compete. These economic necessities have led cable operators and cable networks to integrate vertically. Major owners of chains of cable systems own at least partial interests in most of the popular cable networks.

From the cable operator's point of view, these ownership interests help to guarantee continued availability of programming. From the cable network's point of view, affiliation with a multiple system operator provides a guaranteed base of subscribers that can be held hostage to threats by competing integrated networks to deny access.

The ownership of program services by MSOs raises two concerns. First, MSOs might discriminate against competing program services. For example, they could refuse carriage, charge higher retail prices for competing services, or provide less favorable channel positions. Second, MSOs might refuse to provide the program services in which they have an interest to competing distribution outlets (such as SMATV operators or MMDS broadcasters).

There are several points to make about these concerns. First, vertical integration is a common phenomenon in American industry and in most cases is motivated by nothing more sinister than a desire to lower costs. Sometimes, as perhaps is the case in the cable industry, the cost savings arise from avoiding the difficulty of negotiating and enforcing long-term contracts with suppliers for products that are risky and whose content may be difficult to specify with precision (Fisher, 1985).

Second, discrimination is costly. A cable system that refuses to carry a program service must give up the additional subscribers that would have been attracted by it. Similarly, to deny a program service to a competing local distribution medium is to lose the profits from that sale. The circumstances under which such discrimination is profitable in the cable television industry are rare (Ordover, Sykes, and Willig, 1985).

Of course, apparent discrimination can spring from benign economic motives and can benefit consumers. An example is the exclusive arrangements that are common in the theatrical and television production and distribution businesses, as when a motion picture distributor is the sole domestic distributor of a particular film. This "discrimination" is motivated by producers' desire to maximize the value of their properties. Because such arrangements have the effect of making marginal productions more profitable, output is enhanced, benefiting consumers. Thus, even if cable MSOs did engage in such practices, it would be necessary to analyze the effects on consumers before condemning the structural conditions that gave rise to them.

There is now a good deal of empirical research regarding the extent to which cable operators with ownership interests in cable networks engaged in exclusion or discrimination. The research refuses the notion that a cable operator with an ownership interest in a basic cable network systematically discriminates against competing basic cable networks.[37] In addition, there is evidence that vertically integrated cable MSOs carry more network programming overall than do nonintegrated MSOs.

Klein (1989) found that vertically integrated MSOs were more likely to carry a network in which they had an ownership interest, but that these systems were also more likely to carry networks in which they had no financial interest than non–vertically integrated MSOs. This is not surprising. One reason cable systems invest in cable networks is to ensure that programming is available to enhance the demand for cable service. Vertically integrated systems were found not to discriminate against networks in which they had no ownership interest. The National Telecommunications and Information Administration (1988) reached similar conclusions. Salinger (1988) focused on premium movie channels and found some inconclusive evidence of discrimination by certain MSOs against competing premium movie channels, but these findings cannot be generalized to other program services, particularly basic services. Discrimination in pricing, which they also investigate, is not a concern for basic services because basic services are bundled together at a single price.

Using a sample of 400 cable systems, Klein examined carriage percentage (the percentage of systems carrying each given network) for 20 basic and 8 premium networks. For each network he compared the carriage percentage among the systems that had an ownership interest in that network with the carriage percentage among the systems with no ownership interest in that network. Systems with ownership interests tend to carry their own networks, but this tendency does not imply that these systems discriminate against networks in which they have no ownership interest. Discrimination may be present if vertically integrated systems are less likely to carry networks in which they have no interest. To test this, Klein compared, for various networks, the carriage percentage of the four most vertically integrated MSOs (TCI, ATC, Viacom, and Cablevision) with the carriage percentage of systems with no network interests. For each network, the carriage percentage was computed for those of the four MSOs that had no ownership interest in that network. On average, vertically integrated MSOs' carriage percentages for basic and pay networks in which they had no interest was about five percentage points higher than the carriage percentage among non–vertically integrated systems for the corresponding networks. Klein (1989, p. 44) concluded that "there is no systematic discrimination by vertically integrated MSOs as a group against networks among the top 28 in which they do not have an ownership interest."

Crandall (1990b) used a different methodology, multiple regression analysis, but his findings strengthened Klein's conclusions. Klein compared average carriage percentages among groups of systems that had been categorized according to whether they were vertically integrated. The averages of two groups could, of course, differ for reasons other than vertical integration. In contrast, Crandall compared the carriage decisions of four separate groups of firms. He used a probit model and data from TCI, the largest MSO. The control group (2,175 systems with no vertical integration into cable networks) was paired with each of three groups of firms with TCI affiliation. The first group was 361 systems that were owned and operated by TCI; the second group was 189 systems in which TCI had a majority ownership share but did not manage the systems; and the third group was 41 systems in which TCI had an interest, but neither a majority interest nor management responsibility for the system.

Twenty-four of the largest 27 basic cable networks were analyzed. Crandall (1990b) ranked them by their total basic subscribers.[38] Each network had its own variable to indicate carriage. The variable was assigned a value of one for each system that carried that network and zero for each system that did not carry that network. These dichotomous carriage variables were the dependent variables for 24 separate probit regressions; the explanatory variables in each regression are listed in Table 6.5.

Table 6.6 presents the effect of vertical integration on the likelihood of carriage in Crandall's study. Positive numbers in the column headed "coefficients" indicate that systems owned and operated by TCI had a higher likelihood of carriage than systems that were not vertically integrated. A negative value indicates the opposite: that TCI systems were less likely to carry that network than independent systems.

Table 6.5 Factors taken into account

Cable system variables	Demographic and economic variables
Age of the system	Percentage of county population over age 65
Number of homes passed by the cable	Percentage of county population below age 16
Number of broadcast signals received off-air and carried by the system	Percentage of county population between age 16 and age 20
Miles of cable per basic subscriber	Personal income per household in the county
Monthly price of basic service on the system	Persons per household in the county
Number of basic subscribers on the system	Average employment (employment divided by population)
System channel capacity	
Cable subscribers per home passed	
Date for which data are reported	

Source: Crandall, 1990b, p. 11.

Table 6.6 Effect on network carriage of TCI owned-and-operated systems

Network	Coefficient	t-statistic
With TCI Interest		
CVN	1.86	20.99*
TNT	1.79	20.79*
TDC	1.22	13.10*
QVC	0.83	9.98*
AMC[a]	0.66	6.65*
CNN2	0.32	3.96*
WTBS	0.32	2.56*
CNN	0.24	2.32*
BET	0.19	1.83
No TCI Interest		
LIFE	1.05	11.78*
CSPAN	0.87	10.31*
USA	0.66	6.48*
MTV	0.55	6.44*
TWC	0.45	5.62*
TNN	0.44	4.66*
ESPN	0.26	2.14*
NICK	0.04	0.50
VH1	0.03	0.35
AandE	0.01	0.16
FAM[b]	−0.03	−0.35
HSN	−0.06	−0.41
FNN	−0.26	−2.53*
TLC	−0.47	−3.11*
WGN	−0.42	−5.27*

Source: Crandall, 1990b, p. 14.

* *t*-statistic is significant at 95 percent confidence level.

a. Carriage of AMC may be higher on TCI-affiliated systems because they carry AMC as a basic channel whereas some other systems carry it as a premium channel.

b. The Family Channel is included among the networks with no TCI interest because TCI did not acquire an interest until late 1989.

TCI has an ownership interest in each of the first nine networks listed in Table 6.6. For eight of these networks, TCI owned-and-operated systems are significantly more likely to carry the network than were systems with no network affiliation. In the case of the last network, Black Entertainment Television (BET), there is no statistically significant difference between owned-and-operated systems and non–vertically integrated systems in their carriage decisions.

TCI has no ownership interest in the remaining 15 networks in Table 6.6. For seven of these networks, TCI owned-and-operated systems are significantly more likely to carry the network than the independent systems. Thus, for about half of the networks studied, TCI owned-and-operated systems might be said to "discriminate" in favor of networks in which TCI has no interest. (The term "discrimination" in this context is of course a misnomer.) Only three of the 15 networks in which TCI has no ownership interest are less likely to be carried by TCI owned-and-operated systems than by non–vertically integrated systems. For the remaining five networks, the difference between the two groups of systems is not statistically significant. In sum, TCI does not appear to discriminate against networks in which it has no ownership interest.

This finding has strong implications for other systems. If TCI, one of the largest and most vertically integrated MSOs, does not discriminate, it is unlikely that other vertically integrated MSOs could find it profitable to do so.

Regulatory Policy

If federal, state, or local regulators begin to regulate cable television prices and services on the basis of the cable transmission monopoly and if cable systems are vertically integrated into programming, such regulation will inevitably spill over into regulation of content and raise very substantial First Amendment concerns.[39] If cable regulation takes the form of traditional cost-of-service regulation of a public utility, then all the well-known competitive problems of vertical integration in regulated industries, such as those which led to the dissolution of the Bell System, will arise (Noll and Owen, 1988).

Cable television is unlikely to be given common carrier status in the 1990s. First, the cable industry has probably gained too much political power for such a rule to be imposed over its objections. The industry might come to realize, however, that its interests lie in retention of the local cable monopolies rather than in retaining defensive interests in highly competitive and therefore not very profitable networks. The major MSOs would probably benefit from backing legislation forbidding each of them to own cable networks.

Second, the risk of federal content regulation of cable systems seems, at the moment, remote. Regulatory reform over the past decade has lessened content regulation of broadcasters, despite vigorous congressional pressure to retain the fairness doctrine.

Third, there is no compelling evidence of a pattern of abuse by cable operators that would prompt such regulation. Many homes that could have cable still do not subscribe (about 34 percent of all television households), and cable systems cannot afford to eliminate any program source that would contribute to penetration. Most studies show no pattern of discrimination by vertically integrated systems. Moreover, cable is not as yet, and may never be, the only retail channel of video distribution.

But suppose that cable systems enjoyed significant long-term market power—even that they were "natural monopolies." Does it follow that rate regulation is a desirable policy? In previous decades regulation has been used deliberately to restrict cable offerings. Were the Federal Communications Commission or Congress to impose rate regulation anew or to expand the circumstances in which municipal regulation is permissible, they would risk a substantial degradation of service.

In the past twenty years, policymakers have moved away from rate regulation in a large number of markets. This movement has been fueled in part by a belief that competition allocates resources more efficiently and even more fairly than does regulation, particularly in markets with many rivals.[40] Even in cases where there is a continuing concern over the possible exercise of market power—such as local access/exchange service in the telephone industry—regulators are searching for alternatives to traditional rate-of-return regulation. The FCC itself, for example, has substituted price caps for rate-of-return regulation of AT&T and proposed a similar policy for interstate access charges imposed by local-exchange carriers.[41] Several states have also announced at least a variant of price caps in regulating intrastate telephone rates, and one, Nebraska, has essentially deregulated intrastate telephone service.[42]

The case for abandoning formal rate-of-return regulation of public utilities is well documented.[43] This form of regulation induces inefficient choices between current and capital inputs, reduces the incentive for innovation, and creates incentives to cross-subsidize competitive services from rate-regulated services. Because regulators cannot know the utility's cost function, they cannot regulate the structure of rates efficiently or prevent cross-subsidization.[44] Even if regulators set out to constrain cable prices through some form of regulation other than traditional rate-of-return inquiries, there is some risk that contested regulatory proceedings will lead them or reviewing courts to resort to the traditional tools.

The distorting effects of regulation in public utility markets involve relatively homogeneous services, such as gas, electricity, and voice-grade telecommunications.[45] The effects of regulation on an industry that offers a variety of products or services of different quality would be even more damaging to the public welfare.

Not only would rate regulation of cable television service create confusion over the definition of basic cable service, but it would reduce the incentive to develop new basic services. Such problems may have been unthinkable in the 1960s, when cable was little more than a medium for retransmitting broadcast signals, but are immediately apparent in the 1990s, when cable offers a wide array of differentiated program services whose quality varies substantially. Rate regulation would embroil the industry, its regulators, and, most important, its subscribers in endless disputes over program quality and diversity.

For example, if rates were not permitted to rise with the incremental costs of programming, basic cable networks might be forced to reduce the number of original hours per year or to reduce costs and, therefore, program quality in existing hours. Alternatively, a cable operator might find it necessary to substitute a less attractive basic cable service for a more expensive network. Rate regulation would also create disincentives for the development of new programs or basic cable networks. Would rates be controlled for a given number of basic services? Would cable operators be allowed to increase rates for expanding basic service offerings? Would rate increases be allowed for improvements in any one of the 20 or more current basic cable offerings on a system? If not, would cable operators find it less attractive to underwrite or carry new services or new variants of old services? If so, how would these allowable rate increases be determined?

The FCC's unsuccessful search for a rule that would have allowed it to regulate AT&T's private-line tariffs in the 1960s and 1970s suggests that attempts to regulate highly differentiated cable offerings would prove even more difficult. If these regulatory responsibilities were to be assumed by municipalities, which have far fewer regulatory resources than the FCC has, the result would be even less successful.

The regulatory distortions already mentioned generally occur under attempts to regulate monopoly. If a market is comprised of a number of different sellers, if it is contestable, or if the market is growing rapidly, additional problems arise. Regulation allows incumbent firms to resist entry attempts by arguing that newcomers should not be allowed to price at marginal cost for fear that they will be guilty of "cream-skimming" or that important social values will be sacrificed. Broadcasters persuaded the FCC to hold back the development of cable for many years by arguing that local programs and weaker UHF stations would have to be sacrificed in the wake of the audience diversion created by cable. Such arguments are refuted by the survival of local broadcasting, such as news, weather, and sports, despite cable subscription that has reached more than 50 percent of the nation's television households. In fact, there are far more viable UHF stations than there would be in the absence of cable.

Regulation is also used by firms to oppose rate decreases. In the regulation of trucking and railroads, the Interstate Commerce Commission regularly received far more complaints about rate reductions than about rate increases. Competing long-distance telephone companies regularly oppose AT&T's proposed rate decreases, not its increases. In terms of the current cable market, it is not far-fetched to suggest that new service providers—whether MMDS, SMATV, DBS, or simple overbuilders—could and would oppose attempts by regulated firms to reduce rates when entry threatened or occurred. This stunting of the natural competitive process tends to protect producers, not consumers, and generally reduces economic welfare.

Regulation of basic rates would make it much more difficult for cable operators to position themselves against new competition. New services may be uneconomic if rates are not allowed to vary with the number and quality of service offerings, as they do in an unregulated market. If operators are forced to defend every pricing and service decision in a regulatory forum, resources will be diverted from designing and producing new services to regulatory disputes. Cable systems will find themselves defending rate reductions or realignments in the face of complaints from competitors, not customers, and service quality will surely suffer along with the viability of many existing cable operations. In sum, even if cable systems are thought to have some local market power over subscribers, and even if they charge prices that exceed costs, it is far from clear that cable rate regulation similar to that of public utilities will improve consumer welfare.

Local Broadcasters and Cable

Cable systems face two problems with local broadcasters. First, local broadcasters are concerned that they be carried on cable systems, preferably on the same channel numbers they use for broadcasting, or lower ones. Second, the local broadcasters resent the fact that cable operators do not pay for local signals. Rather than regulating the relationship between cable systems and local broadcasters through legislation such as the 1976 Copyright Act, a market solution would be more efficient. Despite the fact that carriage of local broadcast signals is no longer compulsory, it is doubtful that cable operators wish to stop carrying most local television stations. These stations have popular programs that attract viewers, and their presence on the cable system is important in attracting and retaining subscribers. The majority of cable viewing is of broadcast signals. The importance of stations in this respect varies, however. In the early 1990s network affiliates are probably still essential to the success of a local cable system. So are popular VHF independent stations, particularly those with important local sports broadcast rights. In the absence of the present compulsory license, cable systems might well pay such stations for the right to carry their signals on the cable, just as they pay HBO or CNN.[46] But a new, marginal UHF or low-power station contributes little to the success of a cable system. Cable operators may not carry such stations unless they are paid by the stations to do so, just as marginal basic cable networks sometimes pay cable operators for access.

Even with respect to popular stations, such as network affiliates, there may come a time when their importance to the cable operator is less than cable carriage's importance to the stations. When this happens, the stations will have to pay the cable operators for carriage in order to maintain the audiences that they sell to advertisers, if there is a free market in local copyright transactions. Ultimately, of course, the disadvantages of over-the-air broadcast stations (limited channels, single channels, and no practical means to charge viewers) may lead them to be superseded entirely, either by cable or by some different technology. But not in the 1990s.

In the early 1990s, broadcasters and cable operators have continued their struggle to obtain economic advantage through political intervention. Broadcasters have proposed some legislative reinstatement of the FCC's "must carry" requirement to protect weak independent stations, and also a mechanism whereby cable systems would pay a special tax to compensate broadcasters (including networks) for use of their signals. The cable industry has fought to maintain its compulsory free license for local signals. Few voices have been raised in support of the solution that would probably benefit the public: an unfettered market for program property rights.

Telephone Companies and Cable

When the Bell System was dissolved in 1984, AT&T was supposed to retain all the competitive services, and the regional Bell operating companies (RBOCs) were supposed to be limited to the local exchange telephone business. In practice the RBOCs retained many competitive businesses (intra-LATA toll calling, yellow pages, cellular telephone service, and paging services, for example). In addition, the RBOCs have pressed the courts and Congress for permission to enter more businesses. One of these is cable television.

Although both the judge administering the consent decree under which AT&T was dissolved and Congress itself (in the Cable Communications Policy Act of 1984) have forbidden the RBOCs to offer cable service as such, telephone companies can provide facilities to cable operators who in turn deal with the viewing public. What is wrong with permitting telephone companies to provide cable service? The telephone companies argue that new telephone technology, especially digital transmission and fiber optic cables, make cable service a natural adjunct to the provision of telephone service. Moreover, some cable systems offer services, such as monitoring burglar and fire alarms, that compete with those offered by telephone companies. Telephone companies find it unfair that cable systems can offer such services without being regulated by state utility commissions. Finally, permitting telephone companies to offer cable service may generate some healthy competition, not only in the cable business, but in the telephone business.

Yet precisely because it is a potential threat to local telephone monopolies, telephone companies may seek to control cable service in their own territories. In addition, telephone companies that are subject by utility commissions to binding rate-of-return constraints (as most are) have an economic incentive to expand into and to monopolize related unregulated businesses (Noll and Owen, 1988). Any competition between telephone and cable companies is thus likely to be short-lived, ending in a monopoly by the telephone company, whether or not economically justified.

Despite possible monopolization, there are short-run advantages for the public in such a scenario. First, cable service during the process of monopolization would be cheaper. As penetration rates grew in response to lower prices, there would be further growth in new networks, and therefore more variety for viewers. Second, it seems more than likely that the cable services operated by telephone companies would not be vertically integrated into ownership of program networks or control of content, which would be viewed by the utilities as likely to attract unwanted political attention. (One Bell operating company has proposed providing a service called "video dial tone," akin to the content-neutral transmission service offered for voice messages.) In the absence of vertical integration, First Amendment concerns are sharply reduced.

As a practical matter, telephone company operation of cable services as an integral part of a digital fiber optic local network is many years away. (For a thorough study see Johnson and Reed, 1990.) Any relaxation of the constraint on telephone company ownership of cable business would prompt cross-ownership but not joint production. That is, telephone companies would operate cable systems as separate, independent networks, apart from their local telephone facilities. This would eliminate competition without providing any cost savings at all.

Summary

Cable television systems are local distributors of video products. They compete most directly with other local delivery systems such as television stations, SMATV systems, backyard satellite dishes, video retail stores, and MMDS services. Like other local distributors, their principal long-term strategic problem is to avoid being bypassed by technologies that connect individual homes directly to national program sources. One example of such a threat is direct broadcast satellites.

In contrast to broadcast stations, whose most significant capital assets may be their FCC licenses, cable television systems are very capital intensive. However, cable systems have three advantages in selling video products to consumers. First, they do not suffer from the spectrum limitations affecting all broadcast media, even broadcast satellites. Instead, they can supply an indefinitely large number of channels at decreasing unit cost. Second, they can readily monitor usage of their services, charge user fees, and exclude most of those who do not pay. Third, cable systems obtain their most important service offering, local television broadcasts, free of charge.

Cable television is a business whose existence, structure, and profitability have been determined by government policy. Broadcasters succeeded for a number of years in retarding the development of the cable industry. To do this, the broadcasters persuaded federal regulators to restrict the kinds of services that cable systems could offer. The regulators and legislators who implemented this strategic victory for the broadcasters rationalized their actions by claiming that viewers' interests were being protected. In recent decades the cable industry itself has gained considerable ground in the political arena at the expense of broadcasters, notably by obtaining the right to free use of local broadcast signals. In the early 1990s federal law restricts the ability of regulators at all levels of government—local, state, and federal—to impose economic restrictions on cable.

Communication satellites are key to the success of cable service as a distribution medium. Other than local broadcast signals, all of the offerings of a cable system are received via satellite. Before such satellites were available in the early 1970s, cable service merely improved local television reception and brought in a few distant television stations. Networks that acquire and package programs for sale to cable systems use communication satellites to reach thousands of individual cable operators. But the satellite signals can and do reach individual consumers' backyard antennas, and it is only a cost difference of a relatively few dollars that makes it more economical for these networks to use, rather than to bypass entirely, local cable systems.

Cable television systems have modest economies of scale, both with respect to additional subscribers and with respect to added channels. Because of this, marginal costs are lower than average costs, a strategically dangerous situation when it comes to competitive pricing. Because direct price and service competition between overlapping cable systems (called "overbuilds") is likely to be painful, cable systems rely on local governments to restrict direct cable competition. In most parts of the country cable systems have de facto exclusive franchises. For these reasons it is critically important for cable systems to have good relations with the local political power structure that is in a position to protect their franchises. A positive relationship is accomplished in a variety of ways, including straightforward good performance that makes voters content with the status quo. However, the fact that most cable systems do not face a significant overbuild threat does not mean that they lack competition, especially in the long run.

Cable television's abundance of channels is no competitive advantage unless there is attractive programming available to fill the channels. In an effort to stimulate the growth of cable networks through investment, multiple system operators have integrated backward. Many of the important cable network services are thus owned wholly or in part by MSOs. As we have emphasized, the public-good nature of programming dictates that these program services reach the widest possible potential audience. Any competitive advantage gained by MSOs in denying program services to competing media (MMDS, SMATV, DBS) would be dearly purchased. No doubt for this reason, and despite complaints and policy concerns, there is no empirical evidence that MSOs engage in such a strategy. In any event, there is nothing to prevent the owners of video properties from dealing directly or through other distributors with media that compete with cable systems.

Because cable system owners, including MSOs, cannot effectively exclude competing media, they must seek strategic competitive advantage in the pricing and packaging of their services. Cable systems bundle local transmission services with the program services being transmitted, and they also bundle the program services themselves into packages (''basic service'' and ''premium tiers''). Optimizing the structure and pricing of these bundles is complicated by the provisions of local franchises and the threat of regulation, but it is a marketing challenge crucial to the long-term profitability of individual cable systems. In general, the industry has not yet discovered sophisticated solutions to this important problem.

7 / Advanced Television

Technological change in the delivery of video products is accelerating. Fiber optic cables or even ordinary telephone lines may be used to carry television signals in the 1990s. Direct broadcast satellites and numerous other advances are providing strategic opportunities for the existing industry players and for new players.

It is not our purpose here to predict technological developments. Rather, we explore the economics of a particular set of technical changes that were hotly debated in the late 1980s and early 1990s—"high definition" or "advanced" television signals. The issue of advanced television is worth study, not because it will necessarily revolutionize any part of the industry, but as an illustration of the strategic problems and opportunities created by technological change. Perhaps most noteworthy in this story is how the government and rival industry groups have affected the process by which technical standards are set.

Television picture quality has improved dramatically since the current commercial television service was initiated in the 1940s. The grainy, black-and-white images of the late forties and early fifties have given way to much sharper, color pictures. Audio quality also has improved with advances in stereo sound. Much of the improvement in the quality of television is a product of the gradual technological advancements that are expected as a matter of course in a competitive economy, and in electronics industries in particular. Color and stereo sound, however, are advances that could not have occurred without changes in the NTSC standards that determine the technical characteristics of broadcast television signals. NTSC stands for National Television Standards Committee, a voluntary trade committee that advised the Federal Communications Commission on the design of the first U.S. television broadcast standards. Broadcast television systems that are based on NTSC transmission standards are called NTSC systems.

Why have standards and changes in standards so greatly influenced the evolution of technology and service quality in the television industry? The production of television service requires intricate coordination of numerous agents operating in different locations at different stages of the industry. Standards arise and have economic value because they facilitate the coordination of economic activities. Standards translate what might have been arbitrary choices among conventions and technologies into routines and prerequisites for participation that are taken for granted by all concerned. In this way standards reduce the supervision and consultation among actors that otherwise would be required. In a world where every act was unique and performed only once, standards would serve no useful role. But when acts are repeated and the repeated activities of different economic agents must be coordinated, standards confer tremendous benefits.

Precisely because of the important benefits of standards, they are frequently difficult to modify or replace once established. Indeed, part of their value may arise from their stability. The well-understood benefits of established standards may make economic actors reluctant to experiment with new standards that could be more beneficial. The tendency of standards to be resistant to change means that choices among alternative standards should not be taken lightly. Once established, standards may influence economic activity for decades to come. When industries as large as the global television industry are affected, public interest in standards is justified.

Standards both promote and constrain technical improvement. Most products have value because they can be used with other products that are components of a system of related products used for a common purpose. For example, speakers, turntables, and receivers are all components of stereo systems. Standards promote technical advances by providing guidelines to innovators which, if followed, guarantee that a new product or technique will work with the other components of a product system as a whole. But standards also limit technical improvements by restricting technological options. An important issue in the debate over standards for advanced television is whether radical changes in technical standards are required to achieve the desired improvements in television service. Furthermore, if new standards are necessary for a certain level of technical improvement, will these changes in standards carry costs that outweigh the benefits provided?

Advanced television (ATV) is television capable of providing brighter, sharper images and higher-quality sound than is possible with the technology that is standard today.[1] Although some improvement is possible within the current television system, NTSC standards limit the extent of this improvement. The debate over ATV must therefore be a debate over standards.

Standards affect more than just how products and services are produced. Differences in the technologies and costs associated with different standards may profoundly influence market structures and the strategic opportunities of individual firms. For example, some ATV standards differ in terms of how much spectrum they require for an ATV signal. Because the spectrum for television service is severely limited, the ATV standard chosen may affect the number of over-the-air broadcasters, which may in turn change the number of broadcast networks and the diversity of programming. Standards may also vary according to their compatibility with different transmission media. Thus the choice of standards may affect the competitive positions of broadcast and cable services and of national and international suppliers of television hardware and software.

In this chapter we apply the economist's vision of the standard-setting process to high definition television.[2] We introduce some of the terminology that has become common in the ATV policy debate, provide a brief chronology of significant events in the development of ATV, and look at important technical differences among broadcast ATV systems that have been proposed for implementation in the United States.

Terminology

Numerous improvements over current NTSC service commonly are included under the umbrella of advanced television. Improved definition television (IDTV) represents the most modest of these improvements. IDTV improves picture quality through the use of receiver engineering technologies that reduce picture artifacts (such as ghosts and line crawl) and provide the impression of a richer, denser picture.

Enhanced definition television (EDTV) has better quality than IDTV. This improvement is the product of modest changes in signal processing technology. In most cases, NTSC distribution standards combined with advanced receivers such as those used for IDTV are used for EDTV.

The FCC has defined high definition television (HDTV) to be television service with visual resolution equal to that of 35-millimeter film and sound that meets compact disc standards. The systems now labeled as high definition systems are those that most closely approximate the 35-millimeter ideal; none of them, however, meets this standard. Limitations on channel capacity dictate some compromises. The high definition systems would all require substantial changes from NTSC standards, either by using the spectrum differently or by supplementing NTSC signals with other signals that carry the additional information required for high definition service.

The FCC has no authority to regulate IDTV, because IDTV does not require modifications in the way the radio spectrum is employed. IDTV improvements, therefore, are entirely market driven. Implementation of HDTV and most versions of EDTV for broadcast service require FCC approval.

Brief History

The public broadcaster in Japan, NHK (Japanese Broadcasting Corporation), has sponsored engineering and scientific work on high definition television since the early 1970s, but this work received little attention outside of Japan until NHK announced its intention to begin an HDTV service using a commercially available technology in the 1980s. American and European ATV projects were then initiated in the mid-1980s, spurred by fears that yet another domestic industry would be overwhelmed by superior technology from Japan. NHK began limited HDTV broadcasts with its MUSE direct broadcast satellite service in 1989. Both the European and U.S. efforts remain in the developmental stage in the early 1990s. A consortium of European firms is developing the Eureka HDTV system with considerable support from European governments.

The Eureka system will employ different standards than those employed in Japan, but it is also planned as a DBS service. The

United States is in the process of selecting technical standards for a terrestrial ATV broadcast system. The U.S. focus on terrestrial broadcasting probably reflects the fact that the FCC, which initiated the standards selection process, has clear legal jurisdiction only over transmission media dependent on the electromagnetic spectrum. Thus potential video cassette standards, and probably cable television standards, fall outside of its authority. In addition, the considerably greater economic and political clout of commercial broadcasters in the United States, compared with that of their counterparts in Europe and Japan, has undoubtedly been important in setting the focus on standards for terrestrial, rather than DBS, service.

In November 1987 the Federal Communications Commission appointed the FCC Advisory Committee on Advanced Television Service, headed by Richard Wiley, a former FCC chairman. The committee's mandate is to investigate and report to the FCC on technical, economic, and communications policy aspects of advanced television. The FCC has stated that it does not intend to select an advanced television system until 1993, but in the intervening years it has made decisions that will narrow its options. One of its first acts was to rule out broadcast ATV systems that would preclude, from a technical standpoint, the continued provision of the current (NTSC) over-the-air television service. This ruled out a direct, terrestrial application of the MUSE direct broadcast satellite system used in Japan. The stated rationale for this ruling was to protect consumers' investments in conventional receivers. In the same rulemaking proceeding, the FCC also stated that spectrum used for advanced television would have to come from the heavily used spectrum already set aside for television broadcasting.[3]

Two types of ATV systems have been proposed for adoption in the United States. The first are receiver-compatible systems. They incorporate a NTSC signal within the ATV signal. This is done either by adding information to the NTSC signal within its 6-megahertz television channel or by augmenting the NTSC signal with a second signal that carries additional information that can be received only by ATV receivers.

Second, there are simulcast ATV systems. They use the 6-megahertz television channels currently left vacant to limit interference among NTSC broadcasters. The simulcast systems use advanced engineering techniques to produce signals that (a) carry much more information than do NTSC signals, (b) do not interfere with NTSC broadcasts, and (c) are immune to interference from NTSC broadcasts. Simulcast systems could increase the number of usable television channels and thereby make possible an increase in the number of over-the-air broadcasters. The FCC decided in 1989 that the first HDTV standard would be a simulcast standard. Although ruling out a receiver-compatible, augmentation approach for HDTV, the FCC left open the possibility that it would select standards for an intermediate-quality, receiver-compatible EDTV system.

The Economics of Standard Setting

Research on the economics of standard setting has focused on the ways in which markets select standards, but the roles of public agencies in setting standards has also been an important topic.[4] A particular concern has been the selection of standards when (1) there exist, at least potentially, a variety of technologies that can be employed to accomplish similar objectives, and (2) the value of a technology to each of its users increases with the size of the user base, at least throughout some range.[5] Both of these conditions apply to television.

A restatement of the second condition is that each user of a standard confers benefits on other users of the standard by deciding to use the same technology; each user is said to be the source of a "positive externality." Externalities associated with the joint usage of a technology are referred to as "network externalities." The individual users of a technology are connected in an interdependent network by their reliance on a common set of standards.

For activities characterized by externalities, the usual presumption that competition among self-interested economic agents will promote market-wide efficiency in consumption and production, and thereby most effectively promote social welfare, may not be true.[6] Moreover, standards that best promote social welfare may not be the ones that survive the competitive process. When network externalities are strong, knowledge of demand and supply conditions is not sufficient for predicting the outcome of market competitions among standards. Apparently inconsequential occurrences in the evolution of markets may dramatically affect their ultimate structures and performance.[7]

Sources of Network Externalities

Number of users. If a technology fails to gain adequate acceptance in the market, firms will stop supporting it. Users of the technology will be "orphaned" as replacement parts, repair services, and technical advice dry up. Purchasers of personal computers with CPM operating systems (such as the Osborne), whose manufacturers failed when demand shifted to IBM personal computers and IBM clones with an operating system incompatible with CPM, can testify to the problems of commitments to technologies that fail in the market. Purchasers of Beta-format video cassette recorders are in a similar position. The risks to users associated with the commercial failure of a standard are real, and are greatest when the individual benefits from using a technology depend on the number of users.

System economies. Users may benefit from "system-wide economies" that emerge independently of the activities of any particular user. Many products are components of systems of products and services coordinated by adherence to common technological standards. In many cases the availability and quality of the components of these systems depends on the size of the user base—services and component variety and availability being greater for more widely used technologies. For example, the availability of blacksmith services declined as combustion engines replaced the horse. The dwindling number of blacksmiths, in turn, hastened the relegation of the horse to largely recreational uses.

System economies affect both producers and consumers by lowering production costs. Producers benefit, at least in the short run, from higher profit margins, and consumers benefit because lower costs are reflected in lower prices and better products and services.[8] Like final users of products and services, producers in growing industries may benefit from the emergence of specialized firms or entire industries supplying support services, inputs, and complementary products. The rapid emergence of an independent software industry in response to the growing use of personal computers is a familiar example of the growth of a complementary product industry. Industry costs may also decline if growth makes possible cost savings from increased specialization (Stigler, 1968). Finally, growth may increase publicly financed support for an industry, because publicly funded research by universities and government agencies is more likely to be directed to the problems and challenges of large industries.

Economies of scale. A large user base confers benefits if there are economies of scale. The production of a product or service is characterized by economies of scale if the per unit costs of production decline as the volume of production increases. Producers benefit directly from costs reduced by scale economies, as with system-wide economies, but the benefits of lower costs are also passed on to buyers. The volume of sales and production is limited by the number of users. Therefore, all users may benefit from economies of scale made possible by growth in the user base.

Benefits of experience. Costs may decline or the quality of services and products may improve as a result of experience gained in the manufacture and use of products based on new technologies. To some extent these are benefits conferred on later users of a technology by earlier users (Katz and Shapiro, 1986).[9] Earlier users will benefit also, however, if they adopt new versions based on the same technology or if experience-based cost reductions or product improvements stimulate sales and thus increase the number of users, enhancing other benefits of joint consumption.

Benefits of communication. For some technologies the users constitute a network of individuals able to "communicate" with each other. The benefits of this communication increase with the number of users in the network. The value of a telephone network is greater to each user the more individuals and organizations that can be reached with it. The video cassette recorder is another product with which users of a common standard (VHS) "communicate" with each other. Users benefit from the exchange of cassettes, directly or through video outlets. Equally obvious when identified, but less likely to be recognized because they are so ubiquitous, are the benefits of "communication" through the standardization of items and activities as common as screwdrivers and screw heads, light bulbs and sockets, and the convention that dictates that all drivers stick to a particular side of the road.

Choices among Standards

Timing of introduction. When the value of a technology to its users depends on their number, users' choices among technologies cannot be assumed to be independent. In addition to differences in the technical capabilities of the technologies, individuals choosing among technologies will consider the size of the user base of each and how these user bases grow over time. The interdependence of choices implied by network externalities may make market choices among standards extremely sensitive to small events early in the history of an industry, such as which of the contending technologies is introduced first or which potential users choose first. When choices are interdependent, markets may "lock in" on a particular technology even when better alternatives are available. The following two-user, two-technology example illustrates some of the implications of interdependence of choices in a variety of circumstances.

Suppose there are two technologies, 1 and 2 (developed by firms 1 and 2, respectively), and two potential users of these technologies, user A and user B. Both technologies are network technologies, so the value of either technology to a user is greater if the other user makes the same choice of technology. The benefits to user A of using each of the technologies, depending on whether B chooses the same or the alternative technology, are shown in Table 7.1.[10] Note that the benefits to A of using either technology jointly with B exceed the maximum benefits attainable from using one of the technologies alone. The greater benefits reflect the network externalities associated with using these particular technologies.

The order in which the users select among technologies is important. Suppose B chooses first and A is aware of B's choice. In the

Table 7.1 Benefits to User A of alternative technologies

	User B chooses technology 1	User B chooses technology 2
User A chooses technology 1	12	5
User A chooses technology 2	6	8

Table 7.2 Benefits to User B of alternative technologies

	User A chooses technology 1	User A chooses technology 2
User B chooses technology 1	6	4
User B chooses technology 2	5	7

Table 7.3 Sum of benefits to A and B

	User A chooses technology 1	User A chooses technology 2
User B chooses technology 1	18	9
User B chooses technology 2	11	15

situation depicted, A would always select the same technology as B because the benefits of using B's technology, whichever is chosen, always exceed the benefits of using either technology alone. Thus, B, by moving first, entirely determines A's choice among technologies.

The influence that early adopters may have on later adopters' choices is sometimes referred to as the "bandwagon effect." When the benefits of a large user base are substantial, the achievement by any given technology of a critical mass of early adopters may trigger a rush by other users to the same technology. This may happen even if the bandwagon technology is not the optimal choice for market efficiency. To continue with the example, assume that B's benefits from the two technologies are affected by A's choice as shown in Table 7.2. The total benefits (the sum of benefits to A and to B) associated with the four possible combinations of choices among the technologies are given in Table 7.3.

The total of A's and B's benefits is greatest if both choose technology 1. However, for a variety of reasons, B may select technology 2. Suppose B were unaware of user A. B's options would appear to be a choice between technology 1 with a value of 4 and technology 2 with a value of 5. B would select 2. Even if B knew of A and the consequences for A of B's choosing 2, it would still be in B's self-interest to select technology 2 unless it was possible to arrange beforehand for A to pay B a portion of the increased benefits realized by A if both chose technology 1. Without such a side payment, B, as the first mover, has a choice of benefits of 7 from technology 2 and 6 from technology 1. However, legal and/or regulatory restrictions (such as antitrust laws) may forbid coordinated action. Even if explicit coordination is not forbidden, B still might choose technology 2 out of fear that A might unilaterally commit to technology 1 before an agreement is reached, forcing B to follow suit without a side payment as compensation. Of course, if the roles were reversed and A were choosing first, we would expect technology 1 to be selected. Noncooperative private decisionmaking is most likely to produce the socially preferred choice among technologies if the technology that maximizes social welfare is also the technology of choice for each user individually.

Just as noncooperative decisionmaking in the presence of network externalities may result in a socially incorrect choice among new technologies, so users may become locked in to an old technology even when all would benefit from a new alternative. Although movement to the new technology is likely if each user feels confident that others will follow, doubts that other users will follow may lead each to adopt a wait and see strategy. Use of the old, inefficient technology may then be perpetuated indefinitely. The greater are the benefits of a large user base for the old technology, the more likely this is.

Figure 7.1 illustrates the possibility that a less efficient old technology may be locked in, even when all users would benefit from switching to a new technology.[11] Farrell and Saloner (1985) coined the term "excess inertia" to characterize this type of situation. Unlike the previous example, in this case all users, of which there are N^*, are identical. In the situation portrayed, the new technology has just been introduced, and all users are using the old technology. The current benefits to each of using the old technology are $B_0(N^*)$. Individual benefits could increase to $B_n(N^*)$ if all switched. However, a single user of the new technology would realize benefits of only $B_n(1)$, which is less than $B_0(N^*)$. A total of N' users would have to switch to the new technology before its user benefits equaled what is provided by the old technology. If a single user

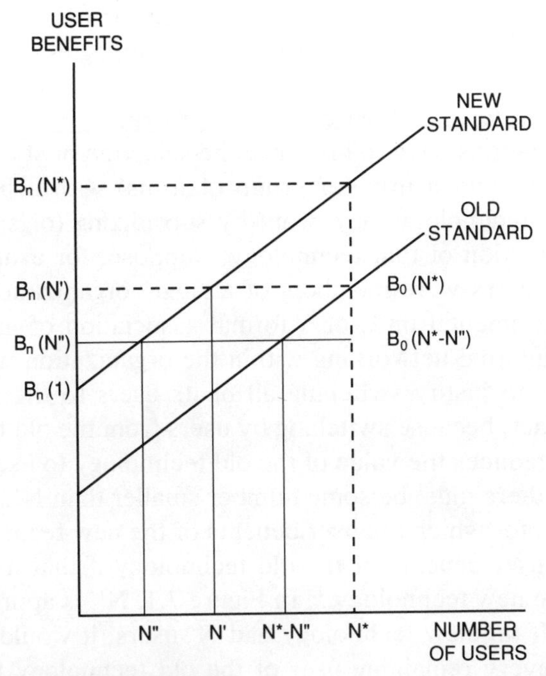

Figure 7.1 Technology "lock-in"

constitutes but a small fraction of N′, then each user may rightfully conclude that a decision on whether to switch or stay with the old technology would not significantly influence others users' perceptions of the merits of switching. That is, the switch by a single user from the old to the new technology would not materially affect the relative positions of the two technology benefit curves. In this situation the temptation to "wait and see" what other users decide is very strong. If all users (or all but a few) follow this strategy, the old technology is effectively locked in (David, 1986).

Differences among users. Old technologies are more likely to be protected by inertia against the challenges of superior new technologies the less confident are individual users that other users share their belief in the superiority of a new technology.[12] Users are more likely to differ in their assessments of the relative merits of alternative technologies if they have different uses for the technologies. Differences among users thus would reinforce the usual fears that one's choice among technologies would not be supported by other users. The uncertainty caused by differences among users may be reduced if the potential adopters of new technologies communicate with each other.

Influential adopters. Figure 7.1 also illustrates the potential that large organizations have to influence the selection of standards and the possibility that a firm (or group of firms) with a proprietary interest in a technology may profit by subsidizing (or sponsoring) the early adoption of that technology. Suppose, for example, that N′ or more users were members of a single organization (a large firm, a governmental unit, or a formal association of users). The benefits of internal networking within the organization would then be sufficient to justify switching all of its users to the new technology. In fact, because switching by users from the old to the new technology reduces the value of the old technology to users that do not switch, there must be some number smaller than N′, which we will call N″, for which the user benefits of the new technology will exceed the user benefits of the old technology if that many users switch to the new technology.[13] In Figure 7.1, N″ is approximately half of N′. If the new technology had N′ users, it would be in the interest of every remaining user of the old technology to switch. An organization with N″ users could switch to the new technology with full confidence that all other users would follow if the organization understood the effects of its switch on the relative benefits of the two technologies to the remaining users. Even if no organization had N″ users, an organization with some significant fraction of N″ users might feel secure in switching if it knew that in doing so it would place one or more other large organizations in the position of being able to switch profitably.[14] Large organizations are more likely to commit early in an attempt to influence their markets' choices among standards than are smaller organizations with no market clout.

Sponsorship. A new technology has a greater chance of displacing an older technology or winning a competition with other new technologies if it is sponsored by a firm with a proprietary interest in its success. In the situation depicted in Figure 7.1, a firm might subsidize the purchases of initial users by making up some or all of the difference between the value of the old technology with N* users [$B_0(N^*)$] and the initial value of the new technology to the first users to switch [$B_n(1)$]. Sponsorship can play an analogous role in competition among new technologies. Sponsored new technologies have an advantage relative to unsponsored new technologies.[15] It might be expected that the amount of sponsorship would reflect the size of the benefits offered by new technologies, so that sponsorship would promote the adoption of the socially optimal choice. But competition among sponsored technologies will not always result in the selection of the technology that maximizes social benefits.[16]

Expectations. In industries in which network externalities are important, expectations may be self-fulfilling. If a user believes that most other users will adopt a particular standard, the user may adopt it, simply to realize the benefits of being part of the large group and to minimize the risks of being stuck with a technology that is not supported in the market. If many expect that a particular standard will eventually dominate and if they act on these expectations, their decisions may trigger a bandwagon for that technology, even if no user prefers that technology over all others.[17] External sources of information, such as the mass media, can initiate such expectations.

Suppliers of complete systems. Sometimes all the components of a new product system cannot be supplied by a single firm. The success of a firm that supplies some of the system components then depends on other firms introducing the remaining parts. In this situation risk-averse firms may choose to delay introducing their own components until the complementary system components are available. If all component suppliers adopt this strategy, however, the new standard will never be introduced. The potentially paralyzing fear of being first to commit to a new standard when the supply of complementary components is uncertain is known as the "chicken-and-egg" problem. Chicken-and-egg risks can be reduced, but not eliminated entirely, if the suppliers of different system components can agree on a schedule for introducing them. There is still a risk that some parties to an agreement will not fulfill their obligation, however. The risk of going first is less likely to deter the introduction of new standards if some members of an industry can provide all the system components. We saw in Chapter 6 one example of this problem: the simultaneous construction of local cable systems and of national networks to supply them with programming.

Standards compatibility. Market choices among standards are not always either-or propositions. Two or more standards frequently coexist within a market, even when it is clear that network externalities are important. Television programs are transmitted and distributed over-the-air, by cable, by microwave, by satellite, and on video cassettes and video disks. People may travel among cities by personal car, bus, plane, and train. Similarly, personal computers based on two incompatible operating systems introduced by Apple coexist with personal computers using two operating systems introduced by IBM and with a variety of other less popular makes based on still different operating systems.

Markets may support multiple, competing standards for several reasons. For some technologies, network externalities may be exhausted with a user base considerably smaller than the entire market. Thus two or more standards may coexist with the users of each realizing maximum network benefits. Even if network benefits continue to increase until all users employ the same technology, multiple technologies may still survive if various subsets of users differ in important ways in the uses they make of the technologies and if the technologies differ in their ability to serve the needs of different users.[18]

Finally, multiple standards may persist if ways can be found to make them sufficiently compatible with each other that users of different standards may share some of the network externalities and system economies made possible by a larger user group.[19] In this case the advantages of specialized technologies may be enjoyed along with the joint consumption benefits from a large group of users (Economides, 1991).

When network externalities create significant inertia for an established technology, a new technology is often introduced in a form that is compatible with the old. This allows users of the new technology to share in some of the network externalities realized by users of the old technology. It may also reduce the risk to firms introducing the new technology by making it possible to introduce components one, or a few, at a time. A common strategy of computer software firms, for example, is to introduce software upgrades that are compatible with older versions.

Firms with proprietary interests in established and dominant technologies may be expected to pursue different compatibility strategies than do firms sponsoring new technologies or technologies targeted to small, specialized niches in markets. Firms with new and niche technologies generally will benefit by making products based on their technologies compatible with dominant technologies. Yet firms with proprietary interests in dominant technologies may try to limit compatibility to saddle small competitors with the disadvantages of small user bases. According to Faulhaber (1987), AT&T's dominant position in the predivestiture telephone industry was derived from its acquisition of the rights to a superior technology for long-distance voice transmission in the early 1900s. AT&T refused to share these rights with competing suppliers of local message service. Independent telephone companies were unable to compete with AT&T because they could not offer customers low-cost connection to other phone users in distant markets. IBM pursued a similar strategy in the market for peripherals to its mainframe computers by denying independent peripheral manufacturers information on the technical specifications of new computers it developed, and by changing interface specifications from model to model.

The Role of Government

"Market failures" are instances in which the market process generates outcomes that differ from the ideal of economic efficiency. The preceding discussion illustrated four significant failures that may occur when markets select standards. First, old technologies may be locked in against the challenges of more efficient new technologies. Second, markets may not select standards that maximize social benefits. Third, too many standards may persist in market equilibrium. Fourth, too few standards may survive the market selection process. In each case an inefficient outcome is the consequence of failure by individual users to reflect in their decision criteria the consequences of their choices for other users and/or actions by users who have only incomplete information about the capabilities of alternative technologies and the preferences of other users.

Government interventions in markets are frequently justified on grounds of externalities and imperfect information. The government has the most power to influence the selection of standards early in the standard-setting process, before technologies have a chance to become locked in. At this early stage, however, the government is likely to be as ignorant of the virtues of various technologies as are private parties, a circumstance David (1987) has described as the "blind giant's" dilemma. Well meant but poorly informed attempts by the government to influence the selection of standards may do more harm than good. Thus any advice concerning government policy for standards selection must be cautionary. Because some of the adverse outcomes of standard setting are more likely if market participants are not fully informed, one role for the government is to ensure that accurate information about alternative standards is disseminated as widely as possible.[20] David suggests that the government may try to slow down the standard selection process in its initial stages, both to ensure that private parties have adequate time to learn about alternative technologies and to ensure that the period during which new standards are introduced is not terminated too quickly by bandwagons for the earliest entrants.

Policymakers might also try to identify situations in which action by a government body to announce a standard or to catalyze market selection of standards is likely to be beneficial. These would include situations in which it is clear that most industry participants desire a new standard and that competing standards promise similar benefits, but no firm or collection of users is willing to bear the risks of being first to adopt. Government selection of a standard may be necessary in cases of extreme inertia, but lesser measures, such as various means of designating official favorites that do not foreclose market choices among alternatives, are often preferable because they initiate private standard selection while preserving some flexibility (Besen and Johnson, 1986).

Direct intervention might be justified when a standard favored by narrow private interests is likely to become dominant but an alternative standard would serve broader social interests. Extremely high standards of evidentiary proof should be required for intervention on these grounds, given that regulators, like private parties, must act on the basis of incomplete information early in the standard-setting process.

Television Standards

There are three sequential stages in the provision of television service: programs must be produced; programs must be delivered to retailers (such as television stations, cable systems, and video stores), and retailers must distribute them to viewers. Each stage can be carried out with a variety of alternative technologies. Programs can be produced live, or they can be prerecorded (on tape or a variety of film formats), or they can combine live and prerecorded segments. Programs can be fed to distributors with microwave relays, by satellite, by fiber or coaxial cables, or by physical transportation of recording media. Retailers use the same technologies, along with the television spectrum, to distribute programs to viewers.

Coordinating Television Service

Standards play two roles in the production of television service. First, they serve as guidelines for coordinating the many activities that take place at each stage. Second, they ensure that the output of one stage is compatible with the technologies and practices employed at the next stage. In the case of distribution, standards ensure that the signals generated are compatible with television receivers.

Most of the discussion about ATV standards has focused on technical standards that govern program production and standards governing the electronic transmission of programs. These standards are reflected in the design of equipment used to produce and distribute programs. We describe production and transmission standards in more detail below.

Although transmission standards govern both the electronic delivery of programs to firms engaged in distribution and the electronic distribution of programs to viewers' receivers, the public debate over transmission standards has focused on standards that will govern over-the-air broadcast transmissions to viewers. This focus reflects an assumption that spectrum limitations will not seriously constrain choices among standards for transmitting programs to distributors, and we will maintain this assumption here. However, tests of alternative ATV systems to be conducted by the FCC's Advisory Committee on Advanced Television Service will include tests of the fidelity of program transmissions to distributors.

Although the relative merits of different production and transmission standards have been heatedly debated, receiver standards have not been a major issue, probably because receiver manufacturers will have no choice but to design and manufacture receivers that are compatible with the transmission/distribution standard(s) that are eventually selected. The debate over transmission standards is thus in effect a debate over receiver standards also. The primary controversy over ATV receivers has centered on William Schreiber's proposal for an open architecture receiver (OAR).[21] Recognizing that ATV receivers will have considerable internal, information-processing capabilities, he proposed that receivers be designed to adapt their display characteristics to conform with the characteristics of signals that might be received through different delivery media. For example, if the cable and broadcast industries used transmission standards with different numbers of horizontal scanning lines, a flexible OAR could accommodate both standards by generating the number of scanning lines appropriate to the signal it was receiving at the moment. Schreiber's open architecture receiver would also be open in the sense that a variety of peripheral devices could be attached to it, similar to the way in which third party peripherals, such as modems and external drives, are connected to personal computers.

The OAR has been vigorously opposed by members of the Electronic Industries Association (EIA), which argues that open architecture would make ATV receivers too expensive for most consumers. The organization also argues that the time required to develop the OAR would needlessly delay the introduction of ATV. The EIA proposes instead that incoming signals from different media be translated by a converter into a common format recognized by the receiver, similar to what is now done with VCRs and cable television. The converter could either sit on top of the receiver or be built in.

Current Standards

Television in the United States today is governed by the NTSC standards for television transmission. These are technical specifications for over-the-air transmission of television signals that were adopted by the FCC when it first allocated spectrum for commercial television service in 1941. Receiver manufacturers subsequently produced television sets designed specifically for the reception of NTSC signals, and programs were produced with techniques adapted to the characteristics of NTSC transmission. Films, which are produced according to standards that preceded television, must be converted (transcoded) to a format suitable for NTSC television transmission.

NTSC is not the only transmission standard currently employed for terrestrial broadcasting. NTSC standards are followed by a number of countries in addition to the United States, but many other countries employ the PAL and SECAM standards that originated in Germany and France, respectively. These three sets of technical standards are incompatible in the sense that transcoding is required before a program produced according to one standard can be broadcast in a country using another. Nevertheless, although there is always some quality loss in transcoding and the process is not cost free, neither quality effects nor transcoding costs have seriously hindered trade in programs between countries governed by different standards.

Technical standards govern many aspects of broadcast television service.[22] All of the standards-governed aspects of television will have to be considered in the choice of ATV standards, but most of the public discussion of standards has focused on a small subset: the number of horizontal lines used to construct television images, picture aspect ratios (the ratio of the width to the height for television pictures), frame rates (the frequency with which images are replaced on the screen), and the amount of spectrum required to transmit the ATV signals. The NTSC service currently enjoyed by American viewers has an aspect ratio of 4:3 and operates at 59.94 frames per second. NTSC utilizes 525 horizontal scanning lines per frame, but considerably fewer than the full complement of horizontal lines is used in constructing NTSC images. The channel assignments awarded to broadcasters by the FCC for NTSC transmissions contain 6 megahertz of spectrum. The PAL and SECAM standards produce pictures with the same aspect ratio as the NTSC standard, but they operate at 50 frames per second with 625 horizontal lines.

Proposed Standards

Almost all of the standards proposed for advanced television service in the United States would produce pictures with greater width to height ratios than those of the NTSC standards. Most have aspect ratios of 16:9, which is close to the motion picture standard, although some are 5:3. The primary argument for higher aspect ratios is that they more closely approximate the shape of the natural human field of vision. The advantages of higher aspect ratios are thought to be greatest for receivers with large screens (greater than 35 to 40 inches, measured diagonally) that would occupy a significant fraction of a viewer's field of vision at normal household viewing distances.

Much of the improvement in picture quality promised by the advanced television systems under consideration by the FCC's Advisory Committee on Advanced Television Service will be accomplished by increasing the number of active scanning lines per frame above what is used for NTSC telecasts.[23] As the size of a television screen is increased, the distance between the lines that make up the picture increases, resulting in the fuzziness and grainy quality that is very apparent with current large-screen receivers (receivers with screens of 40 inches or larger). The loss of resolution with larger screens is offset in ATV receivers by increasing the number of scanning lines used to construct television pictures. With true high definition television, as noted earlier, picture quality would be comparable to that of 35-millimeter film, so that televised images could be shown on cinema size screens without objectionable loss of picture quality.[24]

When ATV service will meet this standard in the United States, if ever, is unclear. Some of the proposed systems increase the number of active scanning lines (those used in the construction of visual images) only slightly over NTSC. Others have over 1000 horizontal scanning lines. The MUSE transmission system already adopted for direct satellite broadcasts by NHK in Japan transmits 1125 lines and 60 fields per second. The Eureka system chosen by most European countries will also be a DBS service and will transmit 1250 lines at 50 frames per second. Europe and Japan also plan to improve their terrestrial television services. Although these will then be superior to current terrestrial service, they will still fall short of the quality that will be provided by the DBS services.

Because the FCC has decided not to add to the spectrum allotted to television service, an ATV broadcast service must be confined to the portion of the spectrum already used by NTSC broadcasters.

Current signals cannot carry enough information for the improvements in picture and sound quality anticipated from EDTV and HDTV. An ATV system can increase the amount of information conveyed by using the current 6-megahertz channels more efficiently or by supplementing NTSC signals with additional audio and visual information carried in augmentation channels of 3 to 6 megahertz. Spectrum or band width available is relatively more abundant for satellites and fiber optic cable and for "hard copy" media such as video cassettes and video disks. Because of differences in technical characteristics and spectrum constraints, transmission standards may vary among the different media.[25]

Until the issue was raised in the context of advanced television, little attention was paid to the distinction between standards that govern television transmission and standards that coordinate the production of television programs. Production standards were developed for use with the transmission systems in different countries and programs traded across national boundaries were transcoded when necessary. New production standards will be required to take full advantage of ATV's potential for improved television service. A number of observers have therefore viewed a worldwide transition to advanced television as an opportunity to establish a single, global production standard to facilitate international program exchanges. Most vocal have been the champions of the 1125 line, 60 frames per second production standard developed by NHK for producing programs to be transmitted by their various MUSE transmission systems. Although the 1125/60 standard is the only ATV production standard currently in use, it still is not used widely and has not been adopted by major studios; in the early 1990s it is being used for the production of some commercials and has seen limited use in producing television programs and films.

Europeans, claiming that the 1125/60 production standard is not very compatible with Eureka transmission standards, have firmly rejected it. The prospects for a global production standard in the near term thus appear to be dim. In addition, NBC, with the support of other major communications companies, has introduced a competing 1050/59.94 production standard, which is said to have greater compatibility with NTSC transmissions.[26] And Zenith has also introduced an HDTV production standard designed for NTSC compatibility. The debate on this issue has made apparent the importance of compatibility among ATV standards for program exchange, both domestically and internationally. We address this concern in the discussion of policy issues later in this chapter.

ATV standards also differ in the amount of the television signal that is encoded and transmitted in digital form. NTSC, PAL, and SECAM all rely on analogue transmission, which means that the information transmitted is represented by distortions in the patterns of the electromagnetic waves employed when broadcasting on a particular television channel. Digitized information is transmitted as a series of ones and zeros.

The MUSE and Eureka direct broadcast satellite HDTV systems also transmit information in analogue form. This contrasts with various hybrid analogue-digital HDTV systems that transmit part of the HDTV signal in analogue form and part of it digitally and with a smaller number of all-digital systems, all of which have been developed in the competition for a U.S. ATV standard since the FCC initiated the ATV standards selection process in 1987. Most of the initial proposals for U.S. ATV standards were for analogue systems. Exceptions were hybrid analogue-digital systems proposed by Zenith and MIT. By 1989, however, the balance of expert opinion had begun to favor the hybrid analogue-digital approach. As a result a number of the early analogue proposals were withdrawn, and some were replaced by newer analogue-digital proposals. Other analogue systems were eliminated by the FCC's decision in favor of the simulcast approach.

By mid-1990 the field of contending systems had been reduced to seven, only two of which were analogue: the proposals by the Japanese Broadcasting Corporation and by the Advanced Television Research Consortium. The ATRC analogue proposal is for an EDTV system that could facilitate the transition to full HDTV. The five remaining proposals are for hybrid analogue-digital systems or all-digital systems. All-digital systems are a recent development that has quickly gained momentum. General Instruments, a late entrant into the HDTV standards race, proposed the first all-digital system (DigiCipher) by paying the testing fees required of system proponents on June 1, 1990, the deadline for payment of these fees set by the Advisory Committee on Advanced Television Service. After DigiCipher's entry into the standards race, Zenith replaced its original analogue-digital system with an all-digital system developed in partnership with AT&T. ATRC also introduced an all-digital system.

With the development of all-digital systems, the United States may have jumped ahead of the Japanese and European HDTV efforts. Both the Japanese and the Europeans started intensive work on HDTV earlier than did companies in the United States and with considerably greater government assistance. Both settled on standards at early dates, before many of the alternative technological options had been explored. Two important advantages of both the hybrid analogue-digital systems and the all-digital systems over the MUSE and Eureka systems are that they are compatible with all television distribution media, including terrestrial broadcasting, and that they substantially reduce interference between adjacent channels, making it possible to use terrestrial channels currently left blank to reduce interference. Thus the digital and analogue-digital systems would make it possible to use more of the spectrum allocated to television than is possible with current television standards.

Because the European nations and Japan plan to preserve terrestrial service, even after their DBS high definition systems are introduced, there may be considerable demand by terrestrial broadcasters in other countries for the technology developed for a terrestrial HDTV system in the United States. The United States may therefore realize long-term benefits from delays in selecting its HDTV standards. This is consistent with David's (1987) observation that the government may beneficially slow down the standards selection process in the early stages of the development of a new technology when the range of its applications and alternative approaches to developing the technology are not well understood.

Technical Compatibility

In its initial notice of inquiry concerning high definition television, the FCC distinguished between two types of compatibility, system compatibility and receiver compatibility: "An ATV system will be considered to be compatible with the existing television channel allotment plan if it operates consistent with the present 6 MHz channelization scheme."[27] That is, an ATV system is system compatible if it does not seriously interfere with the technical capability to continue NTSC television service. System compatibility may also refer to the possibility that two different ATV services, or two different standard television services, may coexist. For example, the current broadcast service and cable television services are system compatible, and the FCC has declared that the next stage in the evolution of broadcast television service must be system compatible with NTSC broadcasting. This limits the threat that an ATV service will render useless the current stock of NTSC receivers.

An ATV system is said to be receiver compatible if its "signal can be decoded and viewed on a conventional NTSC receiver." Again, this definition can be extended to cover the relationships among different ATV systems or among different conventional television systems. The PAL and SECAM broadcast systems, for example, are not receiver compatible with each other or with NTSC. Because cable systems and broadcast systems within a given country use the same receivers, they obviously are receiver compatible.

Receiver compatibility is an application of the broader notion of component compatibility—the extent to which the component activities (production, distribution, and reception) of one television system are compatible with the prior or subsequent activities of another television system. NTSC programs (production signals) are incompatible with PAL and SECAM transmission without transcoding. Similarly, NTSC broadcast signals are incompatible with PAL and SECAM receivers. The discussion that follows focuses primarily on component compatibility, and hence the term "compatibility" refers to any type of component compatibility.

Compatibility is not a simple yes or no proposition, but rather a matter of degree reflecting both cost and aesthetic considerations. At issue is the amount of information lost in converting a program or a transmission designed for one television technology for use with another. The fidelity with which the original source is converted to a second standard can be increased by using more sophisticated and more expensive conversion technologies.

The various media employed for television service in the United States are highly compatible. The same programs are distributed by video cassettes, cable, over-the-air broadcasters, and by satellites to homes equipped with backyard dishes, sometimes called

"TVROs" (television receive-only satellite antennas). Cable and broadcast programs are recorded with VCRs, and cable systems retransmit broadcast signals. This high degree of compatibility among the various television media is to be expected, given that television began as a broadcast service and that the other delivery media were developed as means of enhancing the value of the NTSC receivers already in consumers' homes.

The component compatibility between an ATV and NTSC service can be illustrated diagrammatically (see Figure 7.2). For the purposes of simplification, assume that the various media by which ATV are provided may be treated as a single television service, and make the same assumption for the media by which television service currently is provided. In other words, assume a degree of component compatibility for different ATV media comparable to what now exists for NTSC service. The vertical arrows in Figure 7.2 indicate the types of standards compatibility required to provide each type of service by itself. For each service, the production standard must be compatible with the transmission standard. Simi-

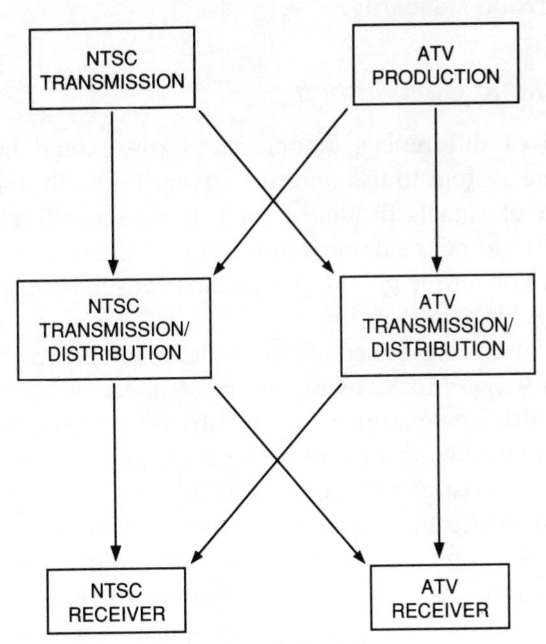

Figure 7.2 Technical compatibility relationships

larly, receivers must be engineered to best exploit the capabilities of the transmission standards in converting signals to pictures and sound. The diagonal arrows depict the various types of compatibility that can exist between the two systems. The arrow connecting NTSC program production to ATV distribution represents the possibility that programs produced according to NTSC production standards will be carried by (or are compatible with) ATV distribution services. Similarly, the arrow from ATV production to NTSC distribution represents their compatibility. The arrow connecting NTSC (ATV) distribution to ATV (NTSC) reception represents the possibility that viewers may be able to view programs carried by NTSC (ATV) distribution firms with ATV (NTSC) receivers.[28] Horizontal arrows connecting the distribution activities of the two systems can be added to illustrate that the transmissions of one system could be picked up and retransmitted by distribution agents of the other service, just as cable systems can retransmit broadcast signals today. An ATV system may be compatible with NTSC service in all, some, or none of the ways illustrated in Figure 7.2, depending on the technical characteristics of the ATV production and transmission standards.

Aesthetic Compatibility

The effects of differences among standards extend beyond the ability of one system to use and to transmit faithfully the information content of signals produced for use with a different system. Compatibility cannot be defined independent of aesthetic concerns. A particular concern is potential problems caused by differences in NTSC and ATV aspect ratios.

These differences may reduce the benefits of shared programs in at least two ways. Most obvious is the "square peg, round hole" problem of fitting a squarer NTSC picture onto a rectangular ATV screen, or reproducing the more rectangular ATV picture on a NTSC screen. Most of the proposed U.S. ATV systems with wide screens plan to use the "letterbox" approach to fitting ATV pictures on NTSC screens. The ATV picture would fill the width of the NTSC screen, with the top and bottom of the NTSC screen blacked out to produce the rectangular shape. Similarly, the sides of an ATV screen could be blacked out for programs with NTSC dimensions. It is probable that viewers will prefer programs that fill their screens over programs that do not. If this is the case, some of the economies that could be realized by using programs from a common source may be lost.

An alternative solution would be to produce programs in the wider ATV aspect ratio and then use NTSC-shaped slices of the actual pictures for NTSC service. This is the manner in which motion pictures are adapted for television viewing. This solution would not rob NTSC viewers of the full use of their screens, but it would force them to accept aesthetic compromises that ATV viewers would not have to make. In any event, the need for programmers to produce programs appealing to NTSC and ATV viewers will almost certainly result in artistic compromises that will prevent the full use of the potential of either type of service. As long as the NTSC audience remains a significant fraction of the total television audience, program producers will probably try to position central characters and significant action closer to the center of the screen than they would if they were producing for an audience viewing only with wide-screen ATV receivers.[29]

Other differences between NTSC service and ATV service may also require aesthetic compromises as long as there are significant audiences for both ATV and NTSC services. Producers may want to emphasize picture details to a greater extent with ATV than with NTSC, for example, to take advantage of higher ATV resolution. It is impossible to know at this early stage to what degree the different aesthetic approaches of the two types of television service will diminish the value of advanced television to consumers.

Network Externalities

If the network externalities of established standards are sufficiently large, it may not be possible to gain acceptance for new technologies that are incompatible with these standards even if all users would benefit. To examine the network externalities associated with the nearly universal use of NTSC standards and NTSC compatible technologies in the U.S. television industry will help us to understand the extent to which NTSC compatibility may be required to successfully introduce ATV in the United States.

The individual benefits of using a technology may increase with the number of users for at least five reasons, as we saw in our discussion of the economics of standards setting. These network externalities contribute to the inertia of established standards, because new technologies start with small user bases and individual benefits are thus more limited, at least initially. NTSC television service exhibits all five externalities.

First, a large user group increases the likelihood of continued commercial support for most new technologies. ATV is not distinguished in this regard. Like many new technologies, early ATV adopters will bear the brunt of the risk that the population of adopters may not achieve the critical mass required to ensure the continued provision of technical support, replacement parts, and complementary products. For business investors, the threat is that inadequate demand will prevent them from ever recovering the cost of their investment. At risk are investments in programs, training and equipment for program production, the research and development costs for ATV hardware, and investments in hardware manufacturing facilities. If ATV does not achieve widespread consumer acceptance, the supply of ATV programming will certainly be limited and could dry up entirely. Early adopters must therefore consider the risk that their investments in ATV receivers and related devices, such as VCRs, will generate limited or no entertainment value. This risk will be greatest if ATV receivers cannot be used for NTSC reception.

Second, if there are system-wide economies, a large user group increases the quality, variety, and availability of system components. In fact, some components may exist only if the user group is sufficiently large. The larger the group of consumers for television programs, the more is spent on the production of individual programs and the greater, and more varied, is the selection. If the installed base of NTSC receivers were divided into two equal but mutually incompatible parts, the quality and variety of programs available for each would diminish. VCRs, stereo television, video games that use television receivers, and other attachments to and enhancements of television sets reflect system-wide economies associated with NTSC standards. At least to some degree, they are all commercial responses to the large market represented by the installed base of NTSC receivers.

System-wide economies are also important in program production. The degree of specialization is limited by the extent of the market (Smith, 1937; Stigler, 1968). In the video industry, thousands of firms provide specialized services that could not be done as well, or at all, if each producer had to be responsible for all of these services. There are firms that specialize in the rental of factors of production: acting talent, production equipment and stages, and special effects expertise. Much of post-production work is also contracted out. The extent to which economies of specialization will be available with an ATV standard when it is new and not widely used is unclear. Some skills and equipment undoubtedly can be adapted to new standards. For others adaptation may not be cost effective.

Third, costs decline with the size of the group if there are economies of scale. Electronic goods—consumer and commercial products—are subject to strong economies of scale in production. Costs and prices of these items often fall dramatically with experience and increasing production volume. The electronic goods required for ATV service will necessarily be produced in small volumes initially, and this will contribute to high introductory prices unless the technology has a sponsor.

The savings realized by spreading the cost of producing a program over a large audience may be the most important source of scale economies in television. For a program of a given budget, the production cost per viewer falls geometrically with the size of the audience. Unless an ATV service is able to share program resources with NTSC services, a high per viewer cost of programming when ATV is new may deter investments in programming. This in turn could deter consumer purchases of ATV receivers.

Fourth, costs may decline and system components may improve with increased experience in producing them. Table 7.4 gives average unit prices for VCRs (calculated as aggregate factory sales divided by the number of units sold) from 1977 to 1986. Prices are reported in nominal and real dollars. A pattern of high introductory prices that fall rapidly for several years is fairly typical of consumer electronics products and many electronic business products as well. Prices for ATV receivers are expected to follow this general pattern. Falling prices reflect reductions in the underlying cost of production. Production costs fall because market growth permits the realization of economies of scale and because costs fall as firms accumulate experience and refine their production techniques over time. High initial costs and prices are a disadvantage to new prod-

Table 7.4 VCR prices, 1977–1986

Year	Units (000)	Value of sales (VOS) ($millions)	Average current price (VOS/units)	Price in constant (1982) dollars
1976	30	—	—	—
1977	250	180	720	990
1978	415	325	883	1018
1979	488	388	795	968
1980	802	620	773	867
1981	1,471	1,126	765	799
1982	2,020	1,303	645	645
1983	4,127	2,162	524	513
1984	7,881	3,585	455	438
1985	11,786	4,738	402	384
1986	13,533	5,257	388	367

Sources: Units and VOS from Darby, 1988, p. 18. Constant dollar prices calculated based on price index for consumer durables.

ucts relative to established products. The higher prices required to cover the higher costs of producing the first units deter the sales that could bring production costs down.

Savvy consumers wait until high introductory prices fall before purchasing a new product. If too many consumers follow this strategy, however, volume may build too slowly to make the introduction of the new product profitable. Introductory prices of ATV receivers are expected to be in the $1,500 to $3,000 range. Most estimates are closer to $3,000. Over time the price should fall toward $1,000.

Competitive firms also strive to improve their products and services over time. The result for electronics products has been spectacular improvements in capabilities and reliability. The growth in sophistication of personal computers is one illustration of this tendency. Another is the evolution of NTSC service. New mixing and editing technologies make possible special effects that were impossible only a few years ago. Similarly, satellite news-gathering technology has made real time coverage of distant events possible for individual broadcast stations, a feat that even the networks could not manage in their coverage of the Vietnam War. These and many other capabilities evolved over time. All of the corresponding technologies for ATV service will not be fully developed when ATV is introduced. Therefore, the initial comparison of the quality of ATV and NTSC service probably will not be as favorable to ATV as the difference in technological potential would suggest.

Fifth, a large user group promotes beneficial interaction and exchanges among users. The trading of video cassettes among VCR owners is perhaps the most obvious example of such a beneficial exchange. Although some direct exchanges among individuals occur, most exchanges are mediated by video rental stores. The fact that the market could not support two incompatible VCR standards (Beta and VHS) is testimony to the importance of standards in coordinating those exchanges.

Of critical importance to the television industry are economies realized through the reuse of programs in syndication and through staggered releases, in different media, which are other forms of market-mediated exchange. After their theatrical releases, films are released on cassettes and then broadcast by cable services, broadcast networks, and independent television stations. Similarly, prime-time network programs may have very long secondary lives in syndication. The ability to tap multiple sources of revenue improves programming.

The infrastructure for distributing NTSC programs has evolved slowly over time for NTSC service. The distribution infrastructure for ATV service will not emerge full blown overnight. ATV programs compatible with NTSC distribution may be advisable.

Obstacles to ATV Adoption

As we have seen, there are many sources of inertia associated with the NTSC standard. The inertia of an established standard is greater if no single firm is able to supply all of the components required to implement a competing standard. It is most difficult to introduce a new technology in these circumstances if system components have no value independent of the other components.

Nonintegrated Suppliers

In the U.S. television industry there are many firms that supply and distribute programs; none, however, can carry out all of the three essential stages of television service: supply (or production), distribution, and receiver manufacture.[30]

With the proliferation of new delivery media there has also been a commensurate increase in the number of components and the complexity of the average viewer's home video system. NTSC color was introduced in the 1950s, when the average viewer had only a single receiver to replace, but most households in the 1990s have multiple receivers and the majority are hooked up to cable or a VCR, or both. To the extent that a new television standard is not compatible with these common attachments to the television receiver, the task of persuading consumers of the advantages of a new television system is that much greater.

Unfortunately, no fully integrated suppliers of the components of advanced television service are likely to emerge. Zenith is the sole remaining U.S.-owned manufacturer of television receivers. To date, it has shown no interest in acquiring production or distribution properties. Foreign receiver manufacturers are barred by law from owning broadcast stations in the United States, and thus are unlikely to surmount the conflicts between manufacturing interests and broadcast interests.

Lack of Significant Sponsors

In the television industry of the 1990s there are no players with the clout or profit incentives comparable to RCA's in the early 1950s. RCA sponsored the introduction of color television in the United States by using the NBC network, which it owned, to supply hundreds of hours of color programming during the years when only a few homes had color televisions. By comparison, ABC and CBS were slow to support the color standard.[31] The hours of programming supplied by the three networks during the first decade of color broadcasting are shown in Table 7.5.

The two strongest networks, NBC and CBS, could count on an average of more than 90 percent of the television audience in the 1950s. Independent stations were few and had tiny audiences. Cable, where it existed, served only to extend the range of broadcast signals. By programming in color, NBC could offer consumers the option of upgrading to color approximately 45 percent of their first choices in programs. In the 1990s the strongest networks have weekly audience shares of approximately 20 percent, less than half

Table 7.5 Yearly hours of network color broadcasts

Year	NBC	CBS	ABC
1954	68	46	0
1955	216	46	0
1956	486	74	0
1957	647	53	0
1958	668	24	0
1959	725	10	0
1960	1,035	5	0
1961	1,650	0	0
1962	1,910	0	35
1963	2,150	5[a]	120
1964	2,135	4	200
1965	4,000[b]	800	600

Source: Reprinted with permission from Ducey and Fratrik, 1989, table 2; based on data from NBC Color Information Department as reported in *Advertising Age,* November 29, 1965, p. 109.

a. Rounded up from figure of 4.5 reported by Ducey and Fratrick.

b. Estimated by Ducey and Fratrick.

of the figure for the 1950s. Although cable networks in the aggregate now attract a substantial audience, no single cable programming supplier has an aggregate audience of even 3 percent.

In 1952 approximately 33 percent of the viewing population lived within the signal coverage areas of NBC's owned-and-operated stations. Four decades later, even though group ownership rules have been relaxed, the largest network-owned station group reaches only about 24 percent of television households. The largest cable multiple system operator, TCI, reaches only about 20 percent of all television households.

Finally, and perhaps most important, RCA was a major manufacturer of television receivers in the 1950s, and had an incentive to promote color television because it profited from the sale of color receivers. No participant in the current television industry is similarly positioned.

Divergent Interests

The proliferation of delivery media has created a divergence of interests much greater than existed when color television was introduced. This diversity will make agreement on an ATV standard more difficult to achieve. Delivery firms in each medium are concerned that the standards selected for an ATV broadcast system will place them at a competitive disadvantage in terms of cost or quality. Representatives of the delivery technologies, however, insist that each delivery medium be allowed to develop standards that allow it to reach its full potential for ATV.

NTSC-ATV Compatibility

NTSC compatibility would make the introduction of ATV easier and its commercial success more likely. The four technical compatibility relationships we discuss here do not directly affect economies of scale and the benefits of experience as sources of inertia. Technical compatibilities, however, can somewhat offset the barriers posed by other sources of inertia. In so doing, they promote increased sales of ATV equipment and thereby hasten the achievement of scale economies and the acquisition of experience.

Production-Transmission Relationships

ATV production to NTSC transmission. ATV programs will have to be converted to a lower-quality format suitable for NTSC transmission if they are to be shared with NTSC television services. There appear to be no serious technical barriers to down-converting ATV programs for this purpose. This type of compatibility reduces the risk that the market for ATV programs will be too small to cover the costs of high-quality ATV productions. Compatibility with NTSC transmission means that from the very beginning, the potential audience for ATV productions will be as large as the number of NTSC-equipped viewers. Furthermore, it ensures a continuing value for ATV programs if ATV does not supplant NTSC.

ATV production to NTSC transmission compatibility could affect viewers and potential providers of ATV transmission services in two ways. On the one hand, if viewers understood that this compatibility reduced the risks of producing ATV programs, they would anticipate more and higher-quality ATV programs than would otherwise be available during the growth phase of ATV. This expectation would encourage the purchase of ATV receivers. Anticipation of a larger ATV audience would encourage more rapid development of ATV transmission capacity and programming. On the other hand, the availability of ATV programs as part of NTSC service (at NTSC quality, of course) would limit any exclusivity value that might help to sell ATV service and hence would retard investment in transmission capacity. There is no apparent reason why ATV programming should be especially distinguished from NTSC programs in terms of content, however; it is doubtful that content distinctiveness will count for much in selling ATV to viewers. ATV production to NTSC transmission compatibility should facilitate the adoption of ATV.

NTSC production to ATV transmission. The facilities for producing made-for-ATV programs probably will be inadequate to supply the program needs of a multichannel ATV service in its early years. As was the case during the transition from black-and-white to color television, the development of full schedules of programs produced for the new standard is likely to take years. Filling in with programs from traditional NTSC sources is desirable. Otherwise, ATV services would be forced to start with very limited broadcast schedules, which would considerably reduce their appeal to viewers. In contrast to down-conversions from ATV to standard television formats, up-conversions from standard formats such as NTSC to ATV standards may entail significant expense and loss of picture quality, depending on how well the technical parameters of the two standards match. Choices among standards therefore greatly affect the degree to which NTSC to ATV compatibility is achieved.

A majority of prime-time dramatic series are produced on 35-millimeter film in the United States. This fact favors ATV; film is a high definition medium, and filmed programs presumably could be easily adapted for ATV distribution just as they are now adapted for NTSC television. For the same reason, the enormous supply of motion pictures filmed in 35 millimeter would facilitate the introduction of ATV. In addition, it should be possible to film many of the programs that are currently recorded on video tape, such as situation comedies, until adequate ATV production facilities are available.

Transmission-Receiver Relationships

ATV transmission to NTSC receiver. To assess the impact of ATV transmission to NTSC receiver compatibility on the prospects for

ATV service, it is necessary to, first, determine the cost to consumers of achieving receiver compatibility and, second, distinguish between ATV systems that would employ the television channels currently used for NTSC broadcasting to carry all or part of the ATV signal and ATV systems (simulcast systems) that would use entirely separate portions of the spectrum. The FCC requires that an ATV system not be technically incompatible with the provision of NTSC service (system compatibility). This requirement will force any ATV system that uses channels currently devoted to NTSC broadcasts to provide the same programs to NTSC viewers that are provided to ATV viewers. The NTSC signal will constitute a portion of the ATV signal. This type of ATV system is, by necessity, receiver compatible. If a receiver-compatible ATV system were adopted and failed commercially, much of the investment in ATV distribution facilities would be lost; there appear to be few alternative uses for the equipment required to enhance NTSC signals.

Investments in ATV distribution systems would not be entirely lost, however, if a simulcast system were adopted and failed. None of the proposed simulcast systems is directly receiver compatible, but their signals may be received on NTSC television sets equipped with decoders. The important question is whether the decoders would be sufficiently inexpensive, and the decoded signals of sufficiently high quality, that consumers would purchase them for access to the additional channels. If viewers equipped their NTSC sets with decoders, the effect would be the same as an increase in the number of NTSC broadcasters, and the ATV licenses and broadcast facilities would retain value as NTSC stations. In the long run, an ATV service that had failed when first introduced could be slowly resurrected if technological advances made high definition receivers, able to receive both ATV and NTSC signals, more affordable. In sum, making ATV transmission compatible with NTSC reception reduces the risks of investing in ATV distribution facilities and thereby promotes the growth of ATV, but only for simulcast ATV systems.

NTSC transmission to ATV receiver. NTSC transmission to ATV receiver compatibility would preserve some of the value of consumers' investments in ATV receivers should ATV be a commercial failure, because the sets consumers purchased could still be used for NTSC service. This type of compatibility would therefore promote ATV receiver sales, which in turn would promote investments in ATV programs and ATV transmission facilities.

If viewers could not use ATV receivers for NTSC service, commitments to ATV would mean either abandoning NTSC service or using different receivers for different television services. Most viewers would likely object to using two receivers, especially if ATV receivers occupy as much floor space as is currently anticipated.[32] If most viewers would choose ATV service over NTSC service, the need to choose between the two could speed the growth of ATV by hastening the demise of the competing service. However, if ATV programming is introduced slowly, as color programming was, then most viewers almost certainly would stick with NTSC if forced to choose between a limited ATV service and current NTSC service. Under these conditions, ATV receivers with the capability to receive NTSC signals would be necessary to the success of an ATV service.

The Importance of Compatibility

The importance of compatibility can be highlighted by examining the consequences of noncompatibility. We first consider the features that an NTSC-incompatible ATV service would have to have to induce consumers to purchase more expensive ATV receivers and give up NTSC service. We then examine the costs and likely revenues of such a service to see if it would provide a reasonable return on investment.

The Perspective of Consumers

Consumers value the production quality of high-budget programs and diversity of program services. The primary sources of high-cost programs, the major television networks, account for a disproportionate share of the total television audience. Growth in the audience shares of cable networks paralleled their move into more expensive programming, a development that further confirms the importance of production quality. Over half of all television households spend $200 to $300 dollars a year and more for the additional channels available over cable, and most of these households also own VCRs—convincing testimony to the value viewers place on a diversity of program sources.

ATV represents a much smaller improvement over NTSC service than the improvement of color over black-and-white service. Even with the dramatic qualitative improvement represented by color, viewers did not rush to buy color receivers compatible with black-and-white service. It is hard to believe, therefore, that viewers would switch from NTSC service to an incompatible ATV service unless the ATV service offered, from the very beginning, several channels with network-quality programming (and network-sized production budgets).

The Perspective of Investors

Viewers have demonstrated by what they are willing to pay for video cassettes and cable that they value the chance to choose among many channels. It seems unlikely that many, if any, viewers would be willing to give up NTSC service for an incompatible ATV service that offered only a few channels. Nevertheless, we will assume, for the sake of argument, that if production values (and program budgets) for ATV programs were comparable to what we now observe for the three major broadcast networks, a two-channel ATV service would attract viewers at the rate forecast in Darby (1988). The rapid growth scenario in Darby's study of the ATV market is probably the most optimistic of the major ATV forecasts.[33]

A reasonable approximation to the annual operating costs of the broadcast networks is revenue net of profits, calculated as revenue on network operations minus pretax profits. The 1989 revenue net of profits for the three major broadcast networks ranged from approximately $2.2 billion for CBS to about $2.4 billion for NBC, with an average of approximately $2.3 billion.[34] If we assume, conservatively, that a single channel of network-quality programming can be produced for about $2 billion a year outside of New York (say, in Atlanta), then the annual operating cost of a two-channel, NTSC-incompatible ATV service would be about $4 billion a year.

The Darby forecast is that ATV receivers will be owned by 1 percent of television households eight years after ATV is introduced. The forecast is based on adoption patterns for other consumer electronic products, such as color television, VCRs, home computers, and audio systems. If we assume that two of the current broadcast networks account for approximately 40 percent of the aggregate television audience and revenues, two advertiser-supported ATV channels capturing all of the ATV audience should realize audience and revenues equal to about 2.5 percent of two current broadcast networks (1 percent divided by 40). The three major broadcast networks averaged slightly over $2.5 billion in revenue apiece on network operations in 1989. If we further assume that the ATV service will have a 1 percent share of viewers for all of its first seven years (rather than building up to that level of penetration over the seven-year period), annual revenue would be about $125 million (.025 times $5 billion).

Given these assumptions, program and operating costs would exceed revenue by $3,875,000,000 each year. At a 10 percent interest rate, the present discounted value of annual losses of this magnitude over the first seven years of the ATV service would be approximately $20.5 billion. ATV penetration would have to reach 40 percent of television households before revenue would begin to cover costs. Under the Darby rapid growth scenario, this would not happen until the fourteenth year after ATV had been introduced, by which time additional billions in losses would have been incurred.

If the costs of equipping two ATV networks and their affiliates are added in, the total sunk costs incurred during the first seven years of ATV service could rise to over $25 billion. NBC has estimated that the equipment costs for a simulcast HDTV service, which would require all new equipment (as would any system completely incompatible with current service), would be over $400 million. Estimates of the costs of equipping network affiliates for HDTV simulcasting vary from less than $8 million (just to pass the network high definition feed on to viewers) to $38 million (to equip major market affiliates to produce local programs in high definition). If the average cost of ATV equipment for a television station is only $10 million, the cost of equipping the more than 200 affiliates of one of the broadcast networks for ATV broadcasting would be more than $2 billion. For a two-network service, the total cost of ATV facilities would be over $5 billion.

An ATV system not compatible with the current NTSC service could not succeed commercially. Private investors would not (and probably could not) finance the $20 to $25 billion in losses that would be incurred during the first seven years of the service, only then to be faced with additional billions in losses until the system was able to cover its operating costs some years later. Operating losses might be reduced somewhat if ATV were introduced as a pay service such as cable, but the difference would not be enough to alter these conclusions. The 1 percent penetration of television households would be equivalent to approximately 2 percent penetration of cable households if ATV were a cable-only service. If 2 percent of the cable industry's approximately $16 billion in annual revenue were subtracted from the $4 billion operating costs for two network-quality program services, the present value of operating losses over the first seven years would still be $19.5 billion.

ATV will have to be at least partially compatible with NTSC so that it can profitably be phased in gradually. This means consumers must not be forced to choose between ATV and NTSC service. ATV receivers compatible with NTSC signals will be required. Furthermore, to cover the costs of high-quality ATV programs, it will be necessary to distribute them to ATV and NTSC audiences. Hence compatibility of ATV programs with NTSC distribution is required. The ability to use NTSC program sources to fill ATV channel time in the beginning would also help, although this type of compatibility is not strictly necessary to the success of ATV service. Broadcast pay-television services such as ON and Select TV succeeded for a time in offering programming only in the evenings.

Most broadcast engineers believe that it is relatively easy to make any high-quality program source usable in a lower-quality distribution channel, and therefore much of this discussion of the need to make ATV production compatible with NTSC transmission may be moot. Nevertheless, this type of compatibility is important enough to the success of an ATV system that it should not be taken for granted.

Adoption Scenarios

The FCC has decided that a first-stage, terrestrial ATV broadcast service must not create technical barriers to continuing broadcast service for the NTSC receivers currently in use. This policy is an attempt to reduce the probability that one particular ATV scenario will occur—the scenario in which consumer investments in NTSC receivers are rendered obsolete by the development of an incompatible ATV service. It is too early to offer specific predictions for the adoption of ATV in the United States. But it is possible to describe several general scenarios that may occur.

At a minimum, an adoption scenario has two components. The first is a predicted end state or market equilibrium. The second is an adoption path or sequence of adoption decisions by users. An adoption path leads from the current situation to the predicted end state. More elaborate scenarios may include forecasts of the rate of adoption. Here we focus on the plausibility of the end states and adoption paths implicit in proposed ATV systems. Adoption scenarios that do not include improved definition television or enhanced definition television proceed directly to high definition television or project a future in which advanced television never becomes established. Adoption scenarios that include IDTV or EDTV fall into three categories: (1) scenarios in which end states with either IDTV or EDTV are preferable to HDTV and are promoted as policy choices; (2) scenarios in which the development of IDTV or EDTV prevents the development of an HDTV system; and (3) scenarios that view IDTV and EDTV as intermediate stages on the path to HDTV. Scenarios of the first type have been precluded, at least for the immediate future, by the FCC's decision to choose a simulcast HDTV system before considering other, lower-quality, ATV possibilities. Scenarios (2) and (3) are still possible because IDTV development does not require FCC approval and because the FCC left open the possibility of nominating an EDTV standard after selecting standards for HDTV. Scenario (1) could reemerge if HDTV is a commercial failure.

IDTV or EDTV Preferred to HDTV

The policy argument for end states with IDTV or EDTV rather than HDTV is that although HDTV may make possible higher-quality service, IDTV and EDTV are less costly and easier to achieve and the cost savings more than offset the benefits of higher quality. This argument cannot be dismissed out of hand. All productive activities are subject to diminishing returns beyond some point, and this tenet is as true for technical improvements in television service as for other goods and services. NTSC pictures can be improved somewhat by building more sophisticated receivers. Further improvements in quality can be achieved with fairly inexpensive modifications to the NTSC system. Service enhancements beyond this point, given the fixed allocation of spectrum, are substantially more expensive. At some point the benefits of additional technical improvements to viewers will not justify the additional cost. There are limits beyond which additional technical improvements in aural fidelity or visual resolution cannot be perceived. The full cost of technical improvements leading to this state cannot be justified. Ideally, television service will be improved just to the point at which the value of additional small improvements has fallen to the level of their cost, but not below. Whether the ideal level of television quality is quality higher or lower than what we associate with HDTV may not be knowable. Yet the possibility that a qualitatively inferior television service may be socially preferred must be acknowledged.

HDTV could be a threat to a socially more desirable EDTV system if HDTV technology developed first, or if HDTV had a powerful sponsor and EDTV did not. Because EDTV is less sophisticated and therefore presumably easier to develop than HDTV, EDTV would be expected to develop first, especially if it promises greater net benefits. A powerful sponsor for HDTV (such as the Japanese government or NHK) could reverse this sequence by accelerating the development of the more sophisticated alternative. However, if it is a foreign government or corporation that is subsidizing the development of HDTV, then it may be reasonable for the United States to view this subsidy as a benefit, because it makes it unnecessary to consume U.S. resources to develop advanced television. Unless there are significant negative indirect consequences flowing from this sponsorship, the comparative rankings of EDTV and HDTV should be reconsidered.

IDTV or EDTV Precludes HDTV

Many of the improvements associated with HDTV are being incorporated in high-end television sets of the early 1990s, producing a noticeable improvement in the quality of NTSC pictures. Such improvements constitute improved definition television, IDTV.

(Further enhancements are possible if NTSC standards are modified slightly, while staying within the 6-megahertz limit on band width, to permit engineering improvements that were not possible when the standards were developed.) EDTV involves improved service requiring more than receiver modifications. If there are benefits to be had from switching from standard NTSC to IDTV or EDTV, the incremental benefits of switching from the intermediate system to HDTV must be less than the benefits of switching directly from standard NTSC to HDTV. Because improvements in NTSC service reduce the benefits of switching to HDTV, some observers now argue that the ongoing and potential improvements within the basic NTSC technical framework, such as those now being incorporated in IDTV receivers, will eventually produce an environment in which more advanced ATV systems will not be able to gain a foothold.[35]

This possibility is illustrated in Figure 7.3. Here, as in Figure 7.1, N* is the number of potential users for any technology. Reflecting the advantages of HDTV assumed in this scenario, the user benefits schedule for HDTV lies above the corresponding

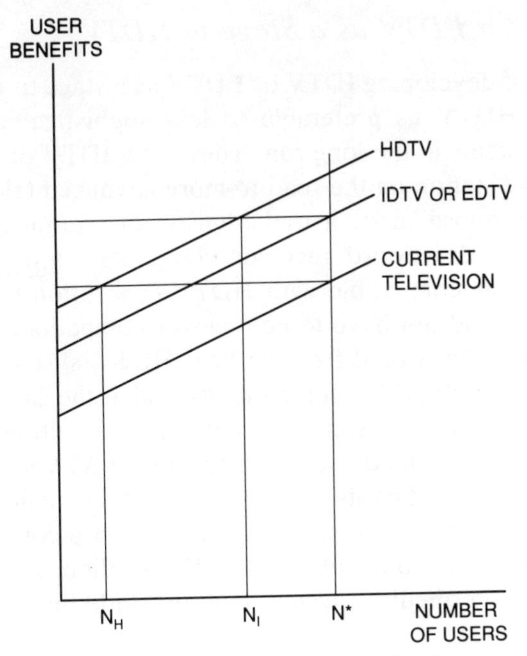

Figure 7.3 Intermediate system inhibits HDTV adoption

schedule for current television (NTSC). The horizontal line through the intersection of the current television benefits line and the vertical line extending up from N* indicate the benefits enjoyed by individual users of the current television system. Following our analysis of NTSC inertia, the starting point (left intercept) of the HDTV benefits schedule lies below the level of benefits provided by the current television system. N_H users would have to switch to HDTV before the user benefits of HDTV would rise to the level currently provided by NTSC service. IDTV, or the development of an EDTV system, would increase the number of viewers (from N_H to N_I in Figure 7.3) that would have to switch to HDTV before individual users realized benefits equivalent to what they enjoyed with the intermediate-quality television system. Viewers and suppliers of system components would naturally be much more reluctant to invest in HDTV under these conditions. This scenario, however, assumes a substantial degree of incompatiblity between an intermediate system and true HDTV. An intermediate system compatible with NTSC and the final HDTV system could facilitate the adoption of HDTV.

IDTV or EDTV as a Stage to HDTV

Proponents of developing IDTV or EDTV as a stage to HDTV tend to see true HDTV as preferable to less sophisticated advanced television systems in the long run. They view IDTV or EDTV as a desirable way station on the road to more advanced television service. They propose, first, a transition to an intermediate EDTV system that is downward receiver compatible with NTSC and upward receiver compatible with HDTV. Transition to an HDTV system that would not have to be receiver compatible with NTSC would follow. This would free the final HDTV system from constraints imposed by NTSC compatibility and at the same time ease the changeover process by making each step a transition to a better, compatible system. Of course, an intermediate ATV system makes no sense if NTSC compatibility does not limit the quality of HDTV service. An intermediate system that is downward compatible with NTSC and upward compatible with HDTV eliminates the possibility that an intermediate system may preclude the attainment of true HDTV.

The two-stage, compatible-systems scenario is illustrated in Figure 7.4. The benefits schedules for incompatible HDTV and intermediate-quality ATV systems are shown as dashed lines.[36] The benefits schedule for an NTSC-compatible intermediate system begins above the level of benefits provided by NTSC when all users are committed to it. Similarly, the intermediate-system-compatible HDTV schedule begins above the maximum level of benefits available with the intermediate system. Therefore, it makes sense for a user to switch to the higher-quality system at each stage.

To summarize: an intermediate-quality ATV system is desirable as a stage leading to full HDTV if (1) compatibility is necessary for HDTV adoption because the benefits realized by the early adopters of an incompatible HDTV system would fall short of the benefits currently available with NTSC *and* (2) NTSC compatibility would limit seriously the capabilities of an HDTV system. If both of these conditions are not satisfied, superior ease and speed of implementation would be the only reason for introducing an intermediate-quality ATV system prior to HDTV if HDTV offered greater bene-

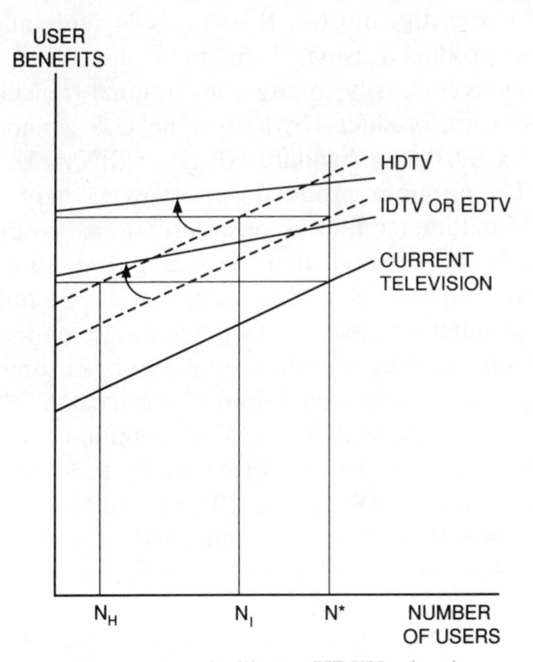

Figure 7.4 Intermediate system facilitates HDTV adoption

fits in the long run. Viewers could then enjoy improved television service while the HDTV system was still being developed. Without these short-term advantages, an intermediate system could only delay unnecessarily the adoption of true HDTV.

Policy Issues

Until recently, television policy in the United States has dealt almost exclusively with domestic issues that related to the sources and content of television programs.[37] By contrast, the public debate over advanced television has focused on international trade issues, and there has been little discussion of how ATV relates to domestic communications policy. We address both of these issues in this section.

International Trade

Films and programs. The adoption of ATV standards throughout the world would affect international trade in television programs and films.[38] Currently, most U.S. exports of films and television programs are produced using 35-millimeter film. This film can be converted relatively easily to any conventional transmission standard for television broadcast, whether the U.S. standard (NTSC) or the various European standards (PAL or SECAM). The advantages of ATV program production, however, may lead to the demise of 35-millimeter film as an international production standard. It has been estimated that producing a one-hour television drama in advanced television video as opposed to 35-millimeter film can reduce production costs by 15 percent. Animation, computer graphics, and many special effects are easier or in some cases only possible with ATV video production. Lower-cost ATV production would be useful even without ATV distribution, because ATV-produced programs can be converted easily to 35-millimeter film for theatrical exhibition (Sugimoto, 1986) as well as to NTSC transmission standards (Rossi and McMann, 1988). Thus there is considerable impetus to move to ATV production before ATV distribution directly to viewers is possible.

A significant problem may arise, however, if the United States and its major trading partners adopt incompatible ATV standards.

See Table 7.6. The leading candidate among production standards is the Japanese-developed 1125/60 standard. Europe almost certainly will adopt the Eureka 1250/50 transmission standard. The Eureka transmission standard would require a production standard that is readily convertible or compatible with 1250/50 transmission. The Japanese 1125/60 production standard is not compatible with 1250/50 transmission. No one has yet attempted to convert programs produced under an 1125/60 standard for distribution for Eureka 1250/50 transmission, and work on this problem is still at an early theoretical stage.[39]

Although it is easy to convert an ATV production standard into any *conventional* transmission standard, it is not easy to translate an ATV production standard into an incompatible *ATV* distribution standard (Wildman, 1991). In other words, one can readily move down (from higher to lower resolution), but not sideways (from one advanced system to another advanced system). If the United States' ATV production standards are incompatible with the ATV distribution standards adopted by its major trading partners, the costs of converting programs produced in the United States could be high.

There may be no economical way to convert U.S.-produced programs to other standards that avoids serious deterioration in the quality of the product. If so, the result of adopting incompatible standards would be to reduce the level of foreign revenues that

Table 7.6 Major television standards

	Location	Production standard	Distribution standard
Conventional			
35-mm film	Worldwide	X	
NTSC (525/59.94)	United States and Japan	X	X
SECAM/PAL (650/50)	Europe	X	X
Advanced (proposed)			
NHK (1125/60)	Japan	X	
Eureka (1250/50)	Europe		X
NBC (1050/59.94)	United States	X	
Zenith (787.5/59.94)	United States	X	X
Muse (various)	Japan		X

U.S.-produced films and television programs earn, as compared with what they could earn with compatible standards. The result would be a needless loss of trade opportunities. Equally serious would be the deterioration in the quality of programs viewed by Americans, because budgets of U.S. productions would fall in response to the diminished ability of U.S. producers to compete for foreign revenues. It is therefore in the interest of the United States to select ATV standards that are compatible with the standards chosen by its trading partners and to minimize the costs of converting programs from one standard to another.

Television Hardware

How will advanced television affect electronics manufacturing in the United States?[40] Some argue that a separate U.S. standard would create manufacturing jobs in the United States if the United States were the sole market for particular ATV equipment.[41] This view is economically unrealistic. Manufacturers of ATV equipment, such as new television receivers, will seek out the methods and locations that minimize manufacturing costs.[42] Currently, it is cheapest to manufacture and assemble most television receivers in the United States. This would be true of advanced television sets irrespective of the standard adopted.

As of 1989, 60 percent of color television sets sold in the United States are produced or assembled in the United States, and 95 percent of 25-inch or larger sets are produced here.[43] Foreign manufacturers of television sets sold in the United States produce many of them in this country. France's Thomson, which owns both the GE and RCA television brand names, has a major production center in the United States. Eight different Japanese firms have television plants in the United States. Output from U.S. plants owned by Japanese companies accounted for 18 percent of Japanese production worldwide in 1986.[44]

A 1988 study done for the Electronic Industries Association by Robert Nathan Associates predicts that 92 percent of ATV sets sold in the United States in the year 2003 will be manufactured in the United States. This prediction is based on the assumption that ATV in the United States will be compatible with existing NTSC standards; it is not based on the assumption that the United States will adopt a unique ATV standard. What is at stake from a point of view of receiver production is the royalty fee representing only about 1.5 percent of the retail price. It is the revenues from this fee that arguably might remain in the United States if ATV television sets were produced with a U.S. design.[45]

Moreover, a unique standard for U.S. ATV does not guarantee that the equipment will be made by U.S. manufacturers.[46] Sony has announced that it will make hardware for different ATV standards if separate standards are adopted in different parts of the world.[47] To protect its consumer electronics industry, Europe has granted proprietary interests in its broadcast standard through patent rights to particular firms. It is difficult to imagine the United States adopting a similar policy, especially given the heavy price that consumers would be forced to pay for such a monopoly.

Domestic Communications Policy

The United States is well along in the process of choosing a new television technology without having seriously considered the structural implications. Fortunately, due more to luck than to foresight, the simulcast technology favored by the FCC for ATV could alleviate many of the structural problems caused by the FCC's poor decisions regarding spectrum allocation in the early 1950s. The spectrum allocation plan favored those who already had VHF licenses. As a result, many markets were served by only three television signals, and UHF stations often had to compete directly with technologically superior VHF stations. The most visible actions by the FCC since that time (promotion of the UHF band and efforts to curb the perceived economic power of the three national broadcast networks) can be seen as attempts to overcome the consequences of those early spectrum allocation decisions.

With NTSC television, a large portion of the spectrum allocated to television does not carry television signals but is used as a buffer to limit interference between the channels that are used. In the VHF band, this means that in each market at least half of the twelve VHF channels must be left blank.[48] In the UHF band, where interference problems are much more severe, eight channels may be left blank for every one used to keep interference to tolerable levels.

Most of the simulcast systems reduce interference between adjacent channels by digitizing all or a portion of the television signal. Digitizing reduces the power required for HDTV broadcasts to as little as 4 percent of what is required for conventional television, and at reduced power, interference is not a problem. At a minimum, this type of simulcast system should double the number of usable broadcast channels in most television markets. It also should reduce the current handicap of UHF television stations. Part of that handicap reflects the greater transmission cost of broadcasting in the UHF band. If simulcast HDTV succeeds, channels currently used for NTSC television could be converted in the long run to HDTV. Even if simulcast HDTV is not commercially successful, HDTV broadcast stations could be used to provide NTSC service if the cost of adapting conventional television sets for reception of ATV signals is not too high and if the signals received in this manner are of acceptable quality.

The past-domination of the television industry by three networks reflected the inability of new networks to find enough stations to carry their signals to U.S. television households. Today there is concern that local television markets may not have enough broadcast stations to constrain effectively the market power of cable systems. By making more channels available, simulcast HDTV may resolve some of these long-standing structural problems that have plagued over-the-air television and cable. Much will depend, however, on the way that new channels are allocated and on plans for eventually converting channels used for NTSC broadcasting to HDTV. Current thinking is that each NTSC licensee will be granted an HDTV simulcast channel in the same market. Although it will be necessary to circumvent current prohibitions on owning two broadcast stations in the same market, this plan has the advantages of facilitating the use of common program sources for HDTV and conventional broadcasting and ensuring that broadcasters will be able to participate in HDTV in the markets they currently serve. If HDTV eventually becomes the dominant service and the NTSC audience continues to shrink, it may be desirable to convert all channels to HDTV service. This transition will force policymakers to make difficult choices among policy objectives that cannot be satisfied simultaneously. These choices can be postponed, but the public interest will be better served if these issues are addressed before HDTV licenses are awarded.

Common ownership of two or more broadcast stations in a single market is now prohibited by law. This prohibition reflects the policy objective of maximizing the number of independent, broadcast gatekeepers in local markets. From this perspective, the more independent gatekeepers there are, the more opportunities there exist for the expression and interplay of contending ideas. The prohibition also reflects deep-seated concerns with the concentration of economic power.

Yet the often-stated objectives of increasing the content diversity of television programming may be better served by allowing dual ownership. Allowing each licensee to operate two channels in a market could counter the natural tendency of competitive, advertiser-supported broadcasters to provide too many channels targeted to mass audiences. A single licensee controlling two channels would be less likely to program both for the same audience than would two independent operators of the same channels. Single-channel competitors would not take into account each channel's effect on the other's audience in deciding on programs, and the audience that one channel would take from another would be viewed as a cost by a broadcaster controlling both.

These ownership issues should be decided before licenses for HDTV channels are awarded. Uncertainty over the property rights implicit in HDTV licenses could retard the development of HDTV services. Because entrenched commercial interests often are able to use the regulatory process to promote their own ends, commercial benefits, once granted, are difficult to revoke.

Summary

Technical standards have shaped the television industry and undoubtedly will play an important role in the evolution of advanced television. As television service with better visual and aural quality than can be provided with the current television system develops, standards provide benchmarks that govern and simplify the coordination of technical relationships in this sophisticated, multistage industry. Standards give the producers and distributors of television programs and the manufacturers of television equipment confidence that what they design and produce can be used with the products of other producers and manufacturers and with the equipment in the hands of consumers. When the benefits of being part of a large group using common standards are significant, products based on new standards may find it difficult to compete. In fact, the handicap of having to start with a small user base may even preclude the successful introduction of technologically superior new standards that would benefit all users. If they are compatible with established standards, new standards have a better chance of success.

The benefits of reliance on NTSC standards in the United States and other NTSC countries are so great that an ATV standard that is not at least partially compatible with the current NTSC service cannot succeed. At a minimum, ATV programs must be compatible with NTSC transmission and reception. An ATV system's chances for success could be further improved if additional types of NTSC compatibilities were built into its standards.

Much attention has been focused on the potential impact of ATV on the international competitiveness of the U.S. consumer electronics industry. It has been suggested that the United States select ATV standards that will favor American electronics manufacturers, an argument that is shaky at best. Electronics manufacturers, regardless of their national affiliation, tend to locate their manufacturing facilities in countries where their products can be produced at least cost. However, the ATV standards that are selected may considerably affect foreign sales of U.S.-produced films and television programs. Foreign sales account for about half of total sales by American studios. A U.S. ATV standard that cannot be converted easily to foreign ATV standards could reduce these sales significantly. The production budgets of American films and programs reflect their sales prospects in other countries. In the long run, American ATV standards incompatible with the ATV systems chosen by major trading partners could lead to lower-budget video productions for U.S. consumption. The ultimate losers from incompatible standards would be American viewers.

The domestic policy implications of ATV should also be addressed before ATV standards are selected and especially before terrestrial broadcast channels are awarded for ATV. The FCC has decided that the new system will provide high definition television rather than enhanced definition television, which is a smaller improvement over current television. The new HDTV system will be a simulcast service that can be provided side by side with the current NTSC service. An advantage of the simulcast approach is that it can greatly increase the number of television channels available for terrestrial television service. Current plans are to award the current holders of NTSC licenses HDTV channels in the same markets so that each local television broadcaster will control two channels in the same market. This plan would give those who now hold television licenses the chance to participate in HDTV in the markets they currently serve. In the longer run, however, a policy that permits a single owner to control two licenses in the same market may not be wise. If HDTV is successful to the point that all television is high definition, then conversion of NTSC channels to HDTV would give television licensees in each local market control of two HDTV channels. Program choice theory tells us that broadcasters will provide viewers with a more diverse menu of programs if each broadcaster programs two channels in a market than if channels are apportioned one to a broadcaster. The political objective of maximizing the number of independent media gatekeepers, however, a priority that stems from First Amendment concerns, would be better served if licensees were restricted to one channel per market. The trade-offs between these two conflicting policy goals should be carefully considered in the design of federal policy for awarding HDTV licenses.

Bibliography

Adams, W. J., and J. L. Yellen. 1976. "Commodity Bundling and the Burden of Monopoly." *Quarterly Journal of Economics* 90:475–498.

Anderson, K. B., and J. R. Woodbury. 1989. "Government Ownership Restrictions and Efficiency: The Case of the FCC's Duopoly Rule." Paper presented at 17th Annual Telecommunications Policy Research Conference, Airlie, Va.

Areeda, P., and D. F. Turner. 1975. "Predatory Pricing and Related Practices under Section 2 of the Sherman Act." *Harvard Law Review* 88:697–733.

Arthur, W. B. 1983. "On Competing Technologies and Historical Small Events: The Dynamics of Choice under Increasing Returns." Working Paper. Institute for Applied Systems Analysis, Laxenburg, Austria.

Barnett, S. R. 1972. "State, Federal, and Local Regulation of Cable Television." *Notre Dame Lawyer* 47:685–814.

Barnow, E. A. 1966–1970. *History of Broadcasting in the United States*. New York: Oxford University Press.

Beebe, J. H. 1972. "Institutional Structure and Program Choices in Television and Cable Television Markets." Research Memorandum no. 131. Research Center for Economic Growth, Stanford University.

———. 1977. "Institutional Structure and Program Choices in Television Markets." *Quarterly Journal of Economics* 91:15–37.

Besen, S. M. 1972. "The Economics of the Cable Television 'Consensus.' " Conference on Communication Policy Research: Papers and Proceedings. U.S. President's Office of Telecommunications Policy, Washington, D.C.

———. 1974. "The Economics of the Cable TV 'Consensus.' " *Journal of Law and Economics*. 17:39–51.

———. 1987. "New Technologies and Intellectual Property: An Economic Analysis." Note N-2601-NSF. May. RAND Corporation, Santa Monica, Calif.

Besen, S. M., and R. Crandall. 1981. "The Deregulation of Cable Television." *Law and Contemporary Problems* 1981:79–124.

Besen, S. M., and L. L. Johnson. 1986. "Compatibility Standards, Competition, and Innovation in the Broadcasting Industry." Prepared for the National Science Foundation. R-3453-NSF. RAND Corporation, Santa Monica, Calif.

Besen, S. M., T. G. Krattenmaker, A. R. Metzger, Jr., and J. R. Woodbury. 1984. *Misregulation Television: Network Dominance and the FCC*. Chicago: University of Chicago Press.

Besen, S. M., and B. M. Mitchell. 1974. Review of Noll, Peck, and McGowan's *Economic Aspects of Television Regulation*. *Bell Journal of Economics and Management Science* 5:301–319.

Besen, S. M., and G. Saloner. 1989. "The Economics of Telecommunications Standards." In *Changing the Rules: Technological Change, International Competition, and Regulation in Communications,* ed. R. Crandall and K. Flamm. Washington, D.C.: Brookings Institution.

Besen, S. M., and R. Soligo. 1973. "The Economics of the Network-Affiliate Relationship in the Television Broadcasting Industry." *American Economic Review* 63:259–268.

Bhagwati, J. 1982. "Directly Unproductive Profit-Seeking Activities." *Journal of Political Economy* 90:988–1002.

Blank, D. M. 1966. "The Quest for Quantity and Diversity in Television Programming." *American Economic Review Papers and Proceedings* 56:448–475.

Blumenthal, M. 1988. "Auctions with Constrained Information: Blind Bidding for Motion Pictures." *Review of Economics and Statistics* 70: 191–198.

Booz, Allen and Hamilton. 1985. "EEC Consumer Electronics—Industrial Policy: Final Report." EEC Information Technologies Task Force, Brussels, Belgium.

Borcherding, T. E. 1978. "Competition, Exclusion, and the Optimal Supply of Public Goods." *Journal of Law and Economics* 21:111–132.

Bowman, G. W., and J. Farley. 1972. "TV Viewing: Application of a Formal Choice Model." *Applied Economics* 4:245–259.

Brennan, T. J. 1990. "Vertical Integration, Monopoly, and the First Amendment." *Journal of Media Economics* 4:57–76.

Brenner, D. 1988a. "Cable Franchising and the First Amendment: Preferred Problems, Undesirable Solutions." *Hastings COM/ENT Law Journal* 10:999–1032.

———. 1988b. "Cable Television and the Freedom of Expression." *Duke Law Journal* 1988:329–388.

Bulow, J. 1982. "Durable Goods Monopolists." *Journal of Political Economy* 90:314–332.

Bulow, J. I., J. D. Geanakopolos, and P. D. Klemperer. 1985. "Multimarket Oligopoly: Strategic Substitutes and Complements." *Journal of Political Economy* 93:488–511.

Carlton, D., and J. Perloff. 1990. *Modern Industrial Organization,* chap. 15. Glenview, Ill.: Scott, Foresman.

Carter, T. B., M. A. Franklin, and J. B. Wright. 1986. *The First Amendment and the Fifth Estate*. Mineola, N.Y.: Foundation Press.

Chamberlin, E. H. 1956. *The Theory of Monopolistic Competition,* 7th ed. Cambridge, Mass.: Harvard University Press.

Clift, C., and A. Greer. 1989. *Broadcast Programming.* Lanham, Md.: University Press of America.

Coase, R. H. 1950. *British Broadcasting: A Study in Monopoly.* Cambridge, Mass.: Harvard University Press.

———. 1959. "The Federal Communications Commission." *Journal of Law and Economics* 2:1–40.

———. 1962. "The Interdepartmental Radio Advisory Committee." *Journal of Law and Economics* 5:17–47.

———. 1972. "Durability and Monopoly." *Journal of Law and Economics* 15:143–149.

Comanor, W. S., and B. S. Mitchell. 1971. "Cable Television and the Impact of Regulation." *Bell Journal of Economics and Management Science* 2:154–212.

Cournot, A. 1963. *The Mathematical Principles of the Theory of Wealth.* Homewood, Ill.: Irwin. Originally published in French in 1838.

Crandall, R. W. 1971. "The Economic Effect of Television-Network Program Ownership." *Journal of Law and Economics* 14:385–412.

———. 1972. "FCC Regulation, Monopsony and Network Television Program Costs." *Bell Journal of Economics and Management Science* 3:483–508.

———. 1974. "The Economic Case for a Fourth Commercial Network." *Public Policy* 12:513–536.

———. 1975. "Postwar Performance of the Motion Picture Industry." *Antitrust Bulletin* 20(1):49–88. Spring.

———. 1990a. "Economic Analysis of Market Structure in the Cable Television Business." Appended to TCI Comments in FCC Mass Media Docket 89-600.

———. 1990b. "Vertical Integration and q Ratios in the Cable Industry." Appended to TCI Reply Comments, FCC Mass Media Docket 89-600.

———. 1990c. "Regulation, Competition, and Cable Performance." Appended to TCI Comments in FCC Mass Media Docket 90-4.

———. 1990d. "Elasticity of Demand for Cable Service and the Effect of Broadcast Signals on Cable Prices." Appended to TCI Reply Comments in FCC Mass Media Docket 90-4.

———. 1990e. "Problems with the New Improved Warehousing Theory." In Joint Economic Appendix to the Reply Comments of ABC, CBS, and NBC in FCC Mass Media Docket 90-162.

———. 1990f. "The Economic Case against the FCC's Television Network Financial Interest and Syndication Rules." With Jt. Comments of Capital Cities/ABC, CBS, and NBC, FCC MM Docket 90-162.

Darby, Larry F. 1988. "Economic Potential of Advanced Television Products." Report for the U.S. National Telecommunications and Information Administration, Washington, D.C.

D'Aspremont, C., J. J. Gabszewicz, and J. F. Thisse. 1979. "On Hotelling's Stability in Competition." *Econometrica* 47:1145–1150.

David, P. A. 1986. "Understanding the Economics of QWERTY: The Necessity of History." In *Economic History and the Modern Economist,* ed. W. N. Parker. Cambridge: Basil Blackwell.

———. 1987. "Some New Standards for the Economics of Standardization in the Information Age." In *Economic Policy and Technological Performance,* ed. P. Dasgupta and P. L. Stoneman. New York: Cambridge University Press.

Demsetz, H. 1970. "The Private Production of Public Goods." *Journal of Law and Economics* 13:293–306.

Dertouzos, J. N., and S. S. Wildman. 1990. "Competitive Effects of Broadcast Signals on Cable." Submitted as an attachment to the Comments of the National Cable Television Association in FCC Mass Media Docket 89-600. March 1.

De Vany, A. S., R. Eckert, C. J. Meyers, D. J. O'Hara, and R. C. Scott. 1969. "A Property System for Market Allocation of the Electromagnetic Spectrum: A Legal-Economic-Engineering Study." *Stanford Law Review* 21:1499–1561.

Diamond, E., N. Sandler, and M. Mueller. 1983. *Communications in Crisis: The First Amendment, Technology, and Deregulation*. Washington D.C.: Cato Institute.

Dominick, J. R., and M. C. Pearce. 1976. "Trends in Network Prime-Time Programming, 1953–74." *Journal of Communication* 26:70–80.

Ducey, R. V., and M. R. Fratrick. 1989. "Broadcasting Industry Response to New Technologies," *Journal of Media Economics* 2:67–86.

Dunnett, P. 1990. *The World Television Industry*. New York: Routledge.

Eaton, B. C., and R. G. Lipsey. 1975. "The Principle of Minimum Differentiation Reconsidered: Some New Developments in the Theory of Spatial Competition." *Review of Economic Studies* 42:27–49.

———. 1976. "The Theory of Spatial Pre-Emption: Location as a Barrier to Entry." Discussion Paper no. 208. Institute for Economic Research, Queen's University, Kingston, Ont.

Economides, N. 1991. "Compatibility and the Creation of Shared Networks." In *Electronic Services Networks,* ed. M. E. Guerin-Calvert and S. S. Wildman. New York: Praeger. Forthcoming.

Epstein, E. J. 1973. *News from Nowhere: Television and the News*. New York: Random House.

Farrell, J., and G. Saloner. 1985. "Standardization, Compatibility, and Innovation." *RAND Journal of Economics* 16:70–83.

Faulhaber, G. R. 1987. *Telecommunications in Turmoil, Technology, and Public Policy*. Cambridge, Mass.: Ballinger Publishing.

FCC. *See* U.S. Federal Communications Commission.

Fisher, F. M. 1985. "The Financial Interest and Syndication Rules in Network Television: Regulatory Fantasy and Reality." In *Antitrust and Regulation: Essays in Memory of John J. McGowan,* ed. F. M. Fisher. Cambridge, Mass.: MIT Press.

Flaherty, J. A. 1987. "Television—The Challenge of the Future." Address to the Society of Motion Picture and Television Engineers Winter Conference, San Francisco, Calif.

Fortnightly Corp. v. United Artists Television, Inc., 392 U.S. 390 (1968).

Fournier, G. M. 1986. "The Determinants of Economic Rents in Television Broadcasting." *Antitrust Bulletin* 31:1045–1066.

Fournier, G. M., and D. L. Martin. 1983. "Does Government Restricted Entry Produce Market Power? New Evidence From the Market for Television Advertising." *Bell Journal of Economics* 14:44–56.

Fowler, M. S., and D. L. Brenner. 1982. "A Marketplace Approach to Broadcast Regulation." *Texas Law Review* 60:207–257.

Ginsburg, D. H. 1979. *Regulation of Broadcasting*. St. Paul, Minn.: West Publishing.

Glenn, W. E., Director of the New York Institute of Technology, Letter to Sherwin H. Becker, Society of Motion Picture and Television Engineers, May 12, 1988.

Glick, I., and S. Levy. 1962. *Living with Television*. Chicago: Aldine.

Goldenson, L. H. 1991. *Beating the Odds*. New York: Charles Scribner's Sons.

Greenberg, E. 1969. "Television Station Profitability and FCC Regulatory Policy." *Journal of Industrial Economics* 10:210–238.

Greenhut, M. L., G. Norman, and C. Hung. 1987. *The Economics of Imperfect Competition: A Spatial Approach*. Cambridge: Cambridge University Press.

Harris, Barry C., and Joseph J. Simons. 1989. "Focusing Market Definition: How Much Substitution Is Necessary?" *Research in Law and Economics* 12:207.

Hazlett, T. W. 1989. "Cabling America: Economic Forces in a Political World." In *Freedom in Broadcasting,* ed. C. G. Veljanovski. London: Institute of Economic Affairs.

———. 1990a. "The Rationality of U.S. Regulation of the Broadcast Spectrum." *Journal of Law and Economics* 33:133–175.

———. 1990b. "Duopolistic Competition in Cable Television." (And "Reply.") *Yale Journal on Regulation* 7:65, 119, 141–148.

———. 1990c. "The Unregulation, Regulation, Deregulation, and Re-

Regulation of Cable Television." Paper presented to the Western Economic Association, San Diego.

———. 1991. "The Demand to Regulate Franchise Monopoly: Evidence from CATV Rate Deregulation in California." *Economic Inquiry* 29:275–296.

Head, S. W., and C. H. Sterling. 1990. *Broadcasting in America: A Survey of Electronic Media,* 6th ed. Boston, Mass.: Houghton Mifflin.

Herring, J. M., and G. C. Gross. 1936. *Telecommunications: Economics and Regulation.* New York: McGraw-Hill.

Home Box Office, Inc. v. Federal Communications Commission, 567 F.2d 9 (D.C.Cir. 1977).

Hoskins, C., and R. Mirus. 1988. "Reasons for the U.S. Dominance of the International Trade in Television Programmes." *Media, Culture, and Society* 10:499–515.

Hoskins, C., R. Mirus, and W. Rozeboom. 1989. "U.S. Television Programs in the International Market: Unfair Pricing?" *Journal of Communication* 39(2):55–75.

Hotelling, H. 1929. "Stability in Competition." *Economic Journal* 34: 41–57.

Hull, B. B. 1990. "An Economics Perspective Ten Years after the NAB Case." *Journal of Media Economics* 3(1):19–36.

ICF, Inc. 1983. "Analysis of the Impacts of Repeal of the Financial Interest and Syndication Rule." Prepared for the Committee for Prudent Deregulation, Washington, D.C.

Innis, H. 1951. *The Bias of Communication.* Toronto: University of Toronto Press.

Johnson, L. L. 1970a. "Cable Television and the Question of Protecting Local Broadcasting." Report R-595-MF. RAND Corporation, Santa Monica, Calif.

———. 1970b. "The Future of Cable Television: Some Problems of Federal Regulation." Memorandum RM-6199-FF. RAND Corporation, Santa Monica, Calif.

Johnson, L. L., and D. R. Castleman. 1991. *Direct Broadcast Satellites: A Competitive Alternative to Cable Television?* Santa Monica, Calif.: RAND Corporation.

Johnson, L. L., and D. Reed. 1990. *Residential Broadband Services by Telephone Companies? Technology, Economics, and Public Policy.* Santa Monica, Calif.: RAND Corporation.

Johnson, N. 1970. *How to Talk Back to Your Television Set.* New York: Bantam Books.

Joskow, P. L., and N. L. Rose. 1989. "The Effects of Economic Regulation." In *The Handbook of Industrial Organization,* ed. Richard Schmalensee and Robert D. Willig. Amsterdam: North Holland Press.

Katz, M. L., and C. Shapiro. 1986. "Technology Adoption in the Presence of Network Externalities." *Journal of Political Economy* 94:822–841.

Klein, B. 1980. "Transaction Cost Determinants of 'Unfair' Contractual Arrangements." *American Economic Review* 70(2):366–362.

———. 1989. "The Competitive Consequences of Vertical Integration in the Cable Industry." Report on behalf of the National Cable Television Association, Washington, D.C.

Krasnow, E. G., and L. D. Longley. 1973. *The Politics of Broadcast Regulation*. New York: St. Martin's Press.

———. 1982. *The Politics of Broadcast Regulation,* 2nd ed. New York: St. Martin's Press.

Kubey, R., and M. Csikszentmihalyi. 1990. *Television and the Quality of Life: How Viewing Shapes Everyday Experience*. Hillsdale, N.J.: Lawrence Erlbaum Associates.

Land, Herman W., and Associates, Inc. 1968. *Television and the Wired City*. Commissioned by National Association of Broadcasters for the use of the President's Task Force on Communication Policy, Washington, D.C.

Lang, K. 1957. "Areas of Radio Preferences: A Preliminary Inquiry." *Journal of Applied Psychology* 41:7–14.

Levin, H. J. 1971a. "Program Duplication, Diversity, and Effective Viewer Choices: Some Empirical Findings." *American Economic Review Papers and Proceedings* 61:81–88.

———. 1971b. *The Invisible Resource: Use and Regulation of the Radio Spectrum*. Baltimore, Md.: Johns Hopkins Press.

Litman, B. R. 1979. "The Television Networks, Competition and Program Diversity." *Journal of Broadcasting* 23:393–409.

———. 1983. "Predicting Success of Theatrical Movies: An Empirical Study." *Journal of Popular Culture* 16:159–175.

Mandese, J. 1989. "Clearing the Network Air." *Marketing and Media Decisions* 24:115–117.

Manishin, G. 1987. "Antitrust and Regulation in Cable Television: Federal Policy at War with Itself." *Cardozo Arts and Entertainment Law Journal* 6:75–100.

———. 1990. "An Antitrust Paradox for the 1990s: Revisiting the Role of the First Amendment in Cable Television." *Cardozo Arts and Entertainment Law Journal* 9:1–14.

McFarland, H. 1988. "Evaluating q as an Alternative to the Rate of Return in Measuring Profitability." *Review of Economics and Statistics* 70:614–622.

McGowan, J. J. 1967. "Competition, Regulation, and Performance in Television Broadcasting." *Washington University Law Quarterly* 499–520.

McLuhan, M. 1964. *Understanding Media*. New York: McGraw-Hill.
Merline, J. W. 1990. "How to Get Better Cable TV at Lower Prices." *Consumers' Research,* May, pp. 10–17.
Midwest Video Corporation v. Federal Communications Commission, et al., 404 U.S. 1014 (1972).
Minasian, J. R. 1969. "The Political Economy of Broadcasting in the 1920s." *Journal of Law and Economics* 12:391–403.
Minow, N. 1964. "The Vast Wasteland," delivered to the National Association of Broadcasters, May 9, 1961. Reprinted in *Equal Time: The Private Broadcaster and the Public Interest*. N. Minow. New York: Atheneum, Macmillan, 1964.
Morrison, S., and C. Winston. 1986. *The Economic Effects of Airline Deregulation*. Washington, D.C.: Brookings Institution.
Nadel, M. S. 1983. "COMCAR: A Marketplace Cable Television Franchise Structure." *Harvard Journal on Legislation* 20:541–573.
Nathan, Robert R., Associates Inc. 1989. "High Definition TV's Potential Economic Impact on Television Manufacturing in the United States." Chap. 4 in "Television Manufacturing in the United States: Economic Contributions—Past, Present and Future." Prepared for the Electronic Industries Association, Washington, D.C. February.
National Association of Theater Owners v. FCC, 420 F.2d 194 (D.C.Cir. 1969).
National Broadcasting Co., Inc. v. United States, 319 U.S. 190 (1942).
Noam, E. M. 1987. "A Public and Private-Choice Model of Broadcasting." *Public Choice* 55:163–187.
Noll, R. G. 1989. "The Economic Theory of Regulation after a Decade of Deregulation: Comments and Discussion." In *Brookings Papers on Economic Activity, Microeconomics 1989,* ed. S. Peltzman, M. N. Baily, and C. Winston. Washington, D.C.: Brookings Institution.
Noll, R. G., and B. M. Owen. 1983. *The Political Economy of Deregulation: Interest Groups in the Regulatory Process*. Washington, D.C.: American Enterprise Institute for Public Policy Research.
———. 1988. "United States v. AT&T: The Economic Issues." In *The Antitrust Revolution,* ed. J. E. Kwoka, Jr., and L. J. White. Glenview, Ill.: Scott, Foresman.
———. 1989. "United States v. AT&T: An Interim Assessment." In *Future Competition in Telecommunications,* ed. S. P. Bradley and J. A. Hausman. Boston: Harvard Business School Press.
Noll, R. G., M. J. Peck, and J. J. McGowan. 1973. *Economic Aspect of Television Regulation*. Washington, D.C.: Brookings Institution.
Novos, I. E., and M. Waldman. 1984. "The Effects of Increased Copyright Protection: An Analytical Approach." *Journal of Political Economy* 84:236–246.

Ordover, J. A., G. Saloner, and S. Salop. 1988. "Equilibrium Vertical Foreclosure." Mimeo.

Ordover, J. A., A. O. Sykes, and R. D. Willig. 1985. "Nonprice Anticompetitive Behavior by Dominant Firms toward the Producers of Complementary Products." In *Antitrust and Regulation: Essays in Memory of John J. McGowan,* ed. F. M. Fisher. Cambridge, Mass.: MIT Press.

Owen, B. M. 1970. "Public Policy and Emerging Technology in the Media." *Public Policy* 18:539–552.

———. 1975. *Economics and Freedom of Expression: Media Structure and the First Amendment.* Cambridge, Mass.: Ballinger Publishing.

———. 1978. "The Economic View of Programming." *Journal of Communication* 28:43–47.

Owen, B. M., J. H. Beebe, and W. G. Manning, Jr. 1974. *Television Economics.* Lexington, Mass.: Lexington Books, D. C. Heath.

Owen, B. M., and R. Braeutigam. 1978. *The Regulation Game: Strategic Use of the Administrative Process.* Cambridge, Mass.: Ballinger.

Owen, B. M., and P. R. Greenhalgh. 1986. "Competitive Policy Considerations in Cable Television Franchising." *Contemporary Policy Issues* 4:69–79.

Pacey, P. L. 1985. "Cable Television in a Less Regulated Market." *Journal of Industrial Economics* 34:81–91.

Park, R. E. 1972a. "Cable Television, UHF Broadcasting, and FCC Regulatory Policy." *Journal of Law and Economics* 15:207–231.

———. 1972b. "Prospects for Cable in the 100 Largest Television Markets." *Bell Journal of Economics and Management Science* 3:130–150.

———. 1972c. "The Exclusivity Provision of the Federal Communications Commission's Cable Television Regulations." Report R-1057-FF/MF. RAND Corporation, Santa Monica, Calif.

———. 1973. "New Television Networks." Report R-1408-MF. RAND Corporation, Santa Monica, Calif.

———. 1975. "New Television Networks." *Bell Journal of Economics* 6:607–620.

Peltzman, S. 1989. "The Economic Theory of Regulation after a Decade of Deregulation." In *Brookings Papers on Economic Activity, Microeconomics 1989,* ed. S. Peltzman, M. N. Baily and C. Winston. Washington, D.C.: Brookings Institution.

Peterman, J. L. 1979. "Differences between the Levels of Spot and Network Television Advertising Rates." Working Paper no. 22. Federal Trade Commission, Bureau of Economics, Washington, D.C.

Poltrack, D. 1983. *Television Marketing.* New York: McGraw-Hill.

Posner, R. A. 1971. "Taxation by Regulation." *Bell Journal of Economics and Management Science* 2:22–50.

Powe, Jr., L. A. 1987. *American Broadcasting and the First Amendment.* Berkeley, Calif.: University of California Press.

Pruvot, M.-J. 1983. "On the Promotion of Film Making in the Community." Working Document 1-504/83. Report to the European Parliament on behalf of the Committee on Youth, Culture, Education, Information, and Sport. July 15.

Public Broadcasting Corporations of the Federal Republic of Germany, Technical Commission of ARD/ZDF, Office of the Technical Commission. 1988. "High-Definition Television." Memorandum. Institut für Rundfunktechnik, Munich.

Renaud, J. L., and B. R. Litman. 1985. "Changing Dynamics of the Overseas Market Place for TV Programming." *Telecommunications Policy* 9:245–261.

Robinson, G. O. 1978. "The Federal Communications Commission: An Essay on Regulatory Watchdogs." *University of Virginia Law Review* 64:169–262.

Rosse, J. N. 1967. "Daily Newspapers, Monopolistic Competition, and Economies of Scale." *American Economic Review Papers and Proceedings* 57:522–533.

———. 1971. "Credible and Incredible Economic Evidence: Reply to Comments in FCC Docket 18110 (Cross Ownership of Newspapers and Television Stations)." Research Memorandum no. 109. Research Center for Economic Growth, Stanford University.

Rossi, J., and R. McMann. 1988. "The 1125 HDTV Production System and Its Relationship to NTSC and HDTV Broadcast Systems." Unpublished paper.

Rothenberg, J. 1962. "Consumer Sovereignty and the Economics of TV Programming." *Studies in Public Communication* 4:45–54.

Salinger, M. 1988. "A Test of Successive Monopoly and Foreclosure Effects: Vertical Integration between Cable Systems and Pay Services." Graduate School of Business, Columbia University. Mimeo.

Samuelson, P. A. 1954. "The Pure Theory of Public Expenditure." *Review of Economics and Statistics* 36:387–389.

———. 1955. "Diagrammatic Exposition of a Theory of Public Expenditure." *Review of Economics and Statistics* 37:350–356.

———. 1958. "Aspects of Public Expenditure Theories." *Review of Economics and Statistics* 40:332–338.

Schiller, H. I. 1969. *Mass Communications and American Empire*. New York: A. M. Kelley.

Shrieves, R. E. 1987. "The Use of Tobin's q." University of Tennessee. Mimeo.

Smiley, A. K. 1990. "Regulation and Competition in Cable Television." *Yale Journal on Regulation* 7:121–139.

Smith, A. 1937. *An Inquiry into the Nature and Causes of the Wealth of Nations*. New York: Modern Library. Originally published in 1776.

Sony Corp. 1988. "HDTV Studio Origination:1125/60 or 1050/59.94?" HDTV Production Series no. 4. Sony Corporation of America, Park Ridge, N.J.

Sony Corp. 1988. "Why a 60-Field HDTV Production System in a 59.94-Field Environment?" HDTV Production Series no. 3. Sony Corporation of America, Park Ridge, N.J.

Spence, A. M. 1976. "Product Selection, Fixed Costs and Monopolistic Competition." *Review of Economic Studies* 43:217–235.

Spence, A. M., and B. M. Owen. 1975, 1977. "Television Programming, Monopolistic Competition and Welfare." *Quarterly Journal of Economics* 91:103–126. (1975 version published as Appendix to chap. 3 of *Economics and Freedom of Expression: Media Structure and the First Amendment*. B. M. Owen. Cambridge, Mass.: Ballinger Publishing.)

Spitzer, M. L. 1986. *Seven Dirty Words and Six Other Stories: Controlling the Content of Print and Broadcast*. New Haven, Conn: Yale University Press.

———. 1991a. "The Constitutionality of Licensing Broadcasters." *New York University Law Review* 64:990–1071.

———. 1991b. "Justifying Minority Preferences in Broadcasting." *Southern California Law Review* 64(2):293–361.

Steiner, P. O. 1952. "Program Patterns and Preferences, and the Workability of Competition in Radio Broadcasting." *Quarterly Journal of Economics* 66:194–223.

———. 1954. "Economic and Regulatory Problems of the Broadcasting Industry." Papers and Proceedings of the 66th Annual Meeting of the American Economic Association. *American Economic Review* 44: 686–87. Abstract.

Stigler, G. J. 1951. "The Division of Labor Is Limited by the Extent of the Market." *Journal of Political Economy* 59:185–193.

———. 1963. "A Note on Block Bundling." In *The Supreme Court Review,* ed. P. B. Kurland. Chicago: University of Chicago Press.

Sugimoto, M. 1986. "The Technical Characteristics of HDTV." Mimeo.

Tannenbaum, P. H. 1985. "Play It Again Sam: Repeated Exposure to Television Programs." In *Selective Exposure to Communication,* ed. D. Zillman and J. Bryant. Hillsdale, N.J.: Lawrence Erlbaum Associates.

Thompson, E. A. 1968. "The Perfectly Competitive Production of Collective Goods." *Review of Economics and Statistics* 50:1–12.

TV and Cable Factbook (annual). Washington, D.C.: Warren Publishing.

U.S. Cable Communications Policy Act of 1984, 98 Stat. 2779 (1984).

U.S. Copyright Act, 90 Stat. 2541 (1976).

U.S. Department of Commerce, National Telecommunications and Information Administration. 1988. *Video Program Distribution and Cable Television: Current Policy Issues and Recommendations.*

U.S. Federal Communications Commission. 1988. Tentative Decision and Further Notice of Inquiry, In the Matter of Advanced Television Systems and Their Impact on the Existing Television Broadcast Service, Mass Media 125. Docket 87-268.

———. 1990. Report in Mass Media Docket 89-600, released July 31, 1990 ("Report to Congress").

U.S. Federal Communications Commission, Network Inquiry Special Staff. 1980a. "The Market for Television Advertising," Preliminary Report.

———. 1980b. *New Television Networks: Entry, Jurisdiction, Ownership, and Regulation*. Report. 2 vols. Washington, D.C.

U.S. Federal Communications Commission, Office of Network Study. 1963. *Television Network Program Procurement,* Part I. Printed as House Report 281, 88th Congress, 2d sess. Washington, D.C.: Government Printing Office.

———. 1965. *Television Network Program Procurement,* Part II. Washington, D.C.: Government Printing Office.

U.S. Federal Communications Commission, Office of Plans and Policy. 1991. "Broadcast Television in a Multichannel Marketplace." June.

———. Working Party 5. 1988. "Economic Factors and Market Penetration: The Working Party 5 Report to the FCC Planning Subcommittee on Advanced Television Service."

U.S. General Accounting Office. 1989. *National Survey of Cable Television Rates and Services.*

U.S. House of Representatives. 1958. Commerce Committee, *Report on Network Broadcasting,* H.R. 1297, 85th Congress, 2d session. (Barrow Report.)

U.S. National Telecommunications and Information Administration. 1988. *Video Program Distribution and Cable Television: Current Policy Issues and Recommendations.* Washington, D.C.: Department of Commerce.

U.S. Office of Technology Assessment. 1986. "Trade in Services: Exports and Foreign Revenues." Special Report, OTA-ITE-316.

U.S. Office of Telecommunications Policy. 1974. Report to the President by the Cabinet Committee on Cable Communications.

U.S. President's Task Force on Communications Policy. 1968. Final Report.

Varis, T. 1984. "The International Flow of Television Programs." *Journal of Communication* 34(1):143–52.

Vogel, H. L. 1986. *Entertainment Industry Economics: A Guide for Financial Analysis*. Cambridge: Cambridge University Press.

Warren-Boulton, F. R. 1990. "Economic Analysis and Policy Implications of the Financial Interest and Syndication Rule." Submitted in FCC Mass Media Docket 90-162 by ICF, Inc., on behalf of the Coalition to Preserve the Financial Interest and Syndication Rule.

Waterman, D. 1985. "Prerecorded Home Video and the Distribution of Theatrical Feature Films." In *Video Media Competition: Regulation, Economics, and Technology*, ed. E. M. Noam. New York: Columbia University Press.

———. 1987. "Electronic Media and the Economics of the First Sale Doctrine." In *Entertainment, Publishing, and the Arts Handbook*, ed. R. Thorne and J. D. Viera. New York: Clark Boardman.

———. 1988. " 'Narrowcasting' on Cable Television: A Program Choice Model." Paper presented at Telecommunications Policy Research Conference, Airlie, Va.

———. 1990. "Diversity, Quality, and Homogenization of Information Products in a Monopolistically Competitive Industry." Unpublished manuscript.

Waterman, D., and A. Grant. 1989. "Narrowcasting on Cable Television: An Empirical Assessment." Unpublished manuscript.

———. 1990. "Multiple Cable Television System Operators and Monopsony Power." Paper presented at Telecommunications Policy Research Conference, Airlie, Va.

Waterman, D., and M. Weiss. 1989, 1990. "The Effects of Vertical Integration between Cable Television Systems and Pay Cable Networks 1988–89." Working Paper M9010, Department of Economics, University of Southern California (rev. July 1990).

Webbink, D. W. 1973. "Regulation, Profits, and Entry in the Television Broadcasting Industry." *Journal of Industrial Economics* 21:167–176.

Webster, J. G. 1986. "The Television Audience: Audience Behavior in the New Media Environment." *Journal of Communication* 36:77–91.

Webster, J. G., and L. W. Lichty. 1991. *Ratings Analysis: Theory and Practice*. Hillsdale, N.J.: Lawrence Erlbaum Associates.

Webster, J. G., and J. J. Wakshlag. 1983. "A Theory of Television Program Choices." *Communication Research* 10:430–446.

Weiss, M., and M. Sirbu. 1988. "Technological Choice in Voluntary Standards Committees: An Empirical Analysis." Paper presented at the 16th Annual Telecommunications Policy Research Conference, Airlie, Va.

Whinston, M. D. 1990. "Tying, Foreclosure, and Exclusion." *American Economic Review* 80:837–859.

White, L. J. 1985. "Antitrust and Video Markets: The Merger of Showtime and The Movie Channel as a Case Study." In *Video Media Competition: Regulation, Economics, and Technology,* ed. E. M. Noam. New York: Columbia University Press.

Whittemore, H. 1990. *CNN: The Inside Story*. Boston: Little, Brown.

Wildman, S. S. 1978. "Vertical Integration in Broadcasting: A Study of Network Owned and Operated Stations." Bureau of Competition, Federal Trade Commission, Washington, D.C.

———. 1991. "Selecting Television Standards for the United States: Implications for Trade in Programs and Motion Pictures." *Journal of Broadcasting and Electronic Media* 35:189–195.

Wildman, S. S., and N. Y. Lee. 1989. "Program Choice in a Broadband Environment." Paper presented at Integrated Broadband Networks Conference, Columbia University.

Wildman, S. S., and B. Owen. 1985. "Program Competition, Diversity, and Multichannel Bundling in the New Video Industry." In *Video Media Competition: Regulation, Economics, and Technology,* ed. E. M. Noam. New York: Columbia University Press.

Wildman, S. S., and S. E. Siwek. 1987. "The Privatization of European Television: Effects on International Markets for Programs." *Columbia Journal of World Business* 22:71–76.

———. 1988. *International Trade in Films and Television Programs*. Cambridge, Mass.: Ballinger Publishing.

———. 1991. "The Economics of Trade in Recorded Media Products in a Multilingual World: Implications for National Media Policies." In *The International Market for Film and Television Programs,* ed. E. M. Noam and J. Millonzi. Norwood, N.J.: Ablex Publishing.

Wiles, P. 1963. "Pilkington and the Theory of Value." *Economic Journal* 73:183–200.

Winston, Clifford, T. M. Corsi, C. M. Grimm, and C. A. Evans. 1990. *The Economic Effects of Surface Freight Deregulation*. Washington, D.C.: Brookings Institution.

Woodbury, J. R., and K. C. Baseman. 1990. "Assessing the Effect of Rate Reregulation on Cable Subscribers." Paper presented at American Enterprise Institute Conference on Network Industries.

Woodbury, J. R., S. M. Besen, and G. M. Fournier. 1983. "The Determinants of Network Television Program Prices: Implicit Contracts, Regulations and Bargaining Power." *Bell Journal of Economics* 14: 351–365.

Notes

1. Introduction

1. Noll and Owen (1983), Hazlett (1990a).
2. For the FCC's current rules and regulations for television, see annual issues of the *Broadcasting/Cablecasting Yearbook.*
3. See R. Rothenberg, "Black Hole in Television," *New York Times,* October 8, 1990, p. D1.
4. The organic statutes of communication regulation were enacted in the days of radio, long before television was a commercial business. On the history of broadcast regulation before 1976, see Herring and Gross (1936), Coase (1959, 1962), Barnow (1966–1970), Minasian (1969), and Hazlett (1990a).
5. In 1952 the Commission had rejected an all-UHF scheme proposed by the Dumont network that would have put current broadcasters and new stations on the same engineering footing. The FCC claimed that an all-UHF-scheme would hurt existing broadcasters. Throughout the 1950s and early 1960s, the Commission instituted in a limited and tentative fashion a policy of "deintermixture," or separation of cities into all-UHF or all-VHF. But struggling for advertisers and viewers, UHF, without FCC protection, was left to languish until 1961. Only then were technical improvements to UHF put in place through a statute forcing television set manufacturers to install UHF tuners under FCC supervision (Krasnow and Longley, 1973, pp. 96–99). More vigorous institution of "deintermixture" would have degraded the signals received by viewers in markets that became all-UHF, at least in the short run, while increasing the overall number of choices available for most viewers. Technical complications could have arisen in adjacent markets. Whether there would have been a net gain in economic welfare can be debated.
6. The first rules forced cable to carry all competing signals and forbade cable from duplicating local programming for 15 days before and after the local broadcast.
7. Cable systems operating in the top-100 television markets could import distant signals only after a hearing showing "such operation would be consistent with the public interest, and particularly the establishment and healthy maintenance of UHF television broadcast

service." See 2 FCC 2nd 725 (1966). "Leapfrogging," or carriage of popular distant signals instead of closer ones, was prohibited, and cable firms were required to produce local programs. See 20 FCC 2nd 201 (1969). Cable systems built in major markets had to have at least 20 channels, two-way transmission capability, and dedicated access channels. See 25 FCC 2nd 38, 40 (1970).

8. The Commission drew up rules for subscription television in the 1960s that were upheld in court in *National Association of Theater Owners v. FCC,* 420 F.2d 194 (D.C.Cir. 1969). The rules barred subscription television from showing any sports programming carried on standard television and restricted feature films to those that were more than three but less than ten years old. Commercials were barred and no more than 90 percent of programming could be sports and movies. Efforts to stretch these rules to cover pay cable television were shot down in *Home Box Office, Inc. v. Federal Communications Commission,* 567 F.2d 9 (D.C.Cir. 1977).
9. For example, the pre-1972 distant signal restrictions on cable were supposedly designed to protect UHF stations. However, the improved reception cable offered local UHF stations more than offset the inroads made by newly available imported VHFs. So the FCC's restrictive carriage requirements actually helped local VHFs at the expense of local UHFs.
10. Report of the President's Task Force on Communications Policy, 1968, p. 9-23. The Rostow group recommended administrative streamlining, more resources for the FCC and said, "in certain areas, we recommend that policy rely more on market forces, and less on regulation, than in the past" (Introduction, p. 8).
11. For hypotheses regarding the process of reform, see Noll and Owen (1983, pp. 155–160), Peltzman (1989), and Noll (1989).
12. See 87 FCC 2d 200 (1981).
13. On deregulation of radio, see 84 FCC 2d 968 (1981); on deregulation of commercial television, 98 FCC 2d 1076 (1984).
14. See 4 Code of Federal Regulations § 73.37 (1981).
15. Federal Communications Commission, In the Matter of the Instructional Television Fixed Service (MDS) Reallocation, 54 Radio Reg. 2d 107 (1983).
16. "Fox Broadcasting Co.: The Birth of a Network?" *Broadcasting,* April 6, 1987, p. 88.
17. *Home Box Office,* supra, n. 9.
18. *Kagan Media Index,* January 21, 1991, p. 6.
19. See *1989 TV and Cable Factbook,* vol. 57, C-316.

2. The Supply of Programming

1. Other factors exist as well. For example, we do not consider possible positive spillovers among channels, such as the possibility that audience awareness of a program generated by its release in one channel may contribute to its audience in subsequent releases in other distribution channels.
2. This assumption is not strictly true for all VCR owners in the real world, but it is a useful assumption here to clarify the interplay of various factors in the design of windowing strategies.
3. Because it is possible to earn money on money (interest) and because people are impatient, a dollar today is worth more than a dollar tomorrow. The rate of interest determines how much future dollars must be discounted relative to current dollars in weighing trade-offs of profits earned in different periods. For example, a dollar deposited in a savings account paying interest of 10 percent per year will grow to $1.10 in a year's time. However, the account is just the bank's promise to pay $1.10 a year from now in exchange for a dollar deposited today. Therefore, the present value of a dollar a year from now is the amount that, if deposited today, will grow to a dollar in a year, which is slightly under 91¢. Savings accounts, bonds, and other financial instruments facilitate exchanges of purchasing power across time periods. As long as financial markets operate efficiently, present value maximization maximizes the amount of purchasing power available at any point in time.
4. See Waterman (1985) for a detailed analysis of windowing for feature films in the mid-1980s. See also Vogel (1986), chap. 3, for an excellent though brief discussion of feature film windows.
5. Waterman's (1987) estimates are for per viewer profit margins in 1984. Table 2.2 depicts a typical release strategy for the late 1980s. The timing of releases may have changed in the intervening years, but the sequence has stayed pretty much the same with the exception of the syndication windows for basic cable and broadcast television.
6. Situation comedies, the most attractive domestic off-network syndication properties, are still leased to television stations first following their network runs.
7. See Tannenbaum (1985) for a discussion of viewers' preferences for previously watched programs.
8. If viewers could duplicate and distribute programs for less than it costs program producers or commercial distributors, copying theoretically could increase the earnings of program owners by reducing costs more

than revenue. See Besen (1987) and Novos and Waldman (1984) on the economics of unauthorized copying.

9. On international video piracy, see Wildman and Siwek (1988, chap. 6).
10. The production costs of programs previously produced are not affected by decisions regarding their distribution. Therefore, prices for these programs will fluctuate with changes in demand and supply conditions subject only to the constraint that the revenue generated be high enough to cover the costs of distribution and residuals paid to talent. The costs of talent and other factors of production place a lower bound on the price at which new programs may be procured. See Besen et al. (1984) for a detailed analysis of bargaining between program suppliers and program services.
11. This startling property of durable goods supply was first demonstrated by Coase (1972). Subsequent work has shown this conclusion to be quite robust. See, for example, Bulow (1982).
12. Waterman (1988) provides an interesting theoretical analysis of the effect of windowing on the selection of program types.
13. Hoskins, Mirus, and Rozeboom (1989) examined the pricing of American programs sold in foreign markets econometrically and rejected the hypothesis that the distributors of American programs were trying to undercut foreign market domestic producers with unreasonably low prices. They found prices in different markets to be explained almost entirely by factors specific to individual national markets that affected competition among suppliers and the degree of monopsony power possessed by domestic purchasers.
14. It could be argued that distribution costs rise somewhat with the size of the production budget. Take the case of off-network programs syndicated to independent television stations. If more stations carry a higher-budget program, which is reasonable if viewer appeal increases with the budget, then distribution costs will rise with the budget as the number of stations carrying it rises. If an off-network program is scheduled by a cable program service such as Turner Network Television (TNT) or USA Network, the number of local outlets (cable systems) carrying the program will not vary with its budget, because programs on the major cable services are almost never preempted.
15. The financial interest rule originally applied only to networks supplying 15 hours or more of programming to their affiliates weekly. Fox was due to cross the 15-hour threshold with its announced plans for expanding its program schedule in 1991, but it persuaded the FCC to waive temporarily the financial interest rule and the syndication rule

(which now prohibits networks from domestic syndication of their own prime-time programs). The FCC then modified the rule to exempt Fox entirely as long as it does not supply more than 15 prime-time hours per week. See *Federal Communications Commission News Release,* April 9, 1991.

16. The syndication demand for prime-time soaps is much stronger overseas. Networks are not allowed to purchase rights in foreign syndication earnings.
17. *Broadcasting Magazine,* May 1, 1989, and April 30, 1990.
18. P. Harris, "MPAA puts '88 b.o. at peak of $4.46-bil.; aud. gets older," *Variety,* February 15–21, 1989, pp. 1, 4. The average budget for films produced by the major American producers was $18.1 million in 1988, down from $20.1 million the previous year. The $9–10 million per hour figure assumes films average two hours in length.
19. This considerably understates the number of negotiating sessions required, because stations probably would consider more than 15 programs before selecting the 15 they broadcast. In addition, bargaining between stations and program suppliers often requires many bargaining sessions in the syndication market. Of course, similar qualifications apply to transactions between networks and program suppliers.
20. Network affiliates form affiliate organizations to negotiate issues of common concern with their networks. This reduces transactions costs further.
21. New information technologies are reducing this advantage. Station representatives select from the commercial-time inventories of many stations to create advertising packages tailored to individual advertisers. This capability will undoubtedly continue to improve as the technologies for electronic data interchange (EDI) are adapted to this business.
22. Crandall (1975) suggests that the motion picture studios may at one time have acted in concert to fix the number of major motion pictures produced in each year.
23. See the 1984 Department of Justice *Merger Guidelines* for detailed discussions of factors that facilitate and make difficult coordination for anticompetitive ends.
24. See "The Powers That Be in Hollywood," *Broadcasting,* July 11, 1988, pp. 43–60; "The Powers That Be in Hollywood: Part Two," *Broadcasting,* August 1, 1988, pp. 35–45.
25. The data underlying these renewal figures are described in some detail in the discussion of risk sharing in Chapter 5.

3. Traditional Models of Program Choice

1. The program choice models presented in this chapter and in Chapter 4 complement the work of audience researchers who study viewer behavior and ratings. See Webster and Lichty (1991).
2. For Newton Minow's famous speech, "The Vast Wasteland," delivered to the National Association of Broadcasters on May 9, 1961, see Minow (1964). See also N. Johnson (1970).
3. Authors who elaborated on Steiner's basic framework in developing models of program choice include Rothenberg (1962), Wiles (1963), Owen, Beebe, and Manning (1974), and Beebe (1977). Noam (1987) extends the analysis to incorporate political influences on program choice, although he relies more directly on the spatial analysis origins of the Steiner framework.
4. The Noam (1987) and Waterman (1990) models of program choice discussed in Chapter 4 are good examples. More generally, see the wide range of topics addressed from a spatial perspective in Greenhut, Norman, and Hung (1987).
5. See Eaton and Lipsey (1975). This result was foreshadowed in the more general but less formal work on monopolistic competition by Chamberlin (1956).
6. A common denominator program is any program that will be watched by different viewer groups. For the viewer preferences we describe, the common denominator program is never assumed to be the first choice of all viewer groups. In general it will be the less-preferred choice.
7. A full enumeration of viewers receiving each choice is omitted here.
8. Consumer surplus is a common measure used in welfare economics. It refers to the difference between the amount a consumer pays for an item and the amount it is worth to the consumer. For total consumer surplus, the assumption is made that individual consumer surpluses can be added. A more general framework should find the outcome that maximizes total (producer plus consumer) surplus, where a producer's surplus is just profit. Economic efficiency requires maximization of total surplus. Total surplus measures are used to compare outcomes in several of the models presented in Chapter 4.
9. Empirical testing of the model's welfare implications requires that program types be empirically measurable and identifiable with viewer preferences. The Beebe model does not attempt to make program types empirically measurable. See the use of industry categories by Levin (1971a) and other empirical studies by Lang (1957), Land (1968), Bowman and Farley (1972), Dominick and Pearce (1976), Litman

(1979), Waterman and Grant (1989), and Wildman and Lee (1989). See also Owen's (1978) critique of empirical measures of diversity.

10. A pure strategy equilibrium means that the broadcaster finds it most profitable to continue showing a profit-maximizing program as long as the other broadcasters continue to show their profit-maximizing programs. The mixed strategy means that each broadcaster finds it most profitable not to reveal its program decision until absolutely necessary, and to calculate probabilities as to which program to offer viewers. This way each broadcaster keeps its competitors guessing about what it will do.
11. An important consideration here is that B must also pay the $400 fixed cost. This is ensured by copyright laws. If A has exclusive rights, then B must purchase a different program for which viewers have identical preferences; hence, the different program is of the same program type, or a perfect substitute.
12. At some price between 4¢ and 10¢, entry can be avoided. A "limit price" of 8¢ is a possibility, because at this price the second broadcaster would be indifferent about entering and splitting the audience. But a price of 10¢ may be sustained if exit is difficult and a price war is anticipated after entry.
13. On pay television and viewer preferences, see Blank (1966) and Noll, Peck, and McGowan (1973). Comanor and Mitchell (1971), Park (1972a), and Pacey (1985) have estimated consumer demand for television using cable television data. Earlier, Lang (1957) attempted measuring radio preferences. These studies provide insights, but their use for developing public policy is subject to limitations (Besen and Mitchell, 1974).

4. Modern Models of Program Choice

1. Most of the models presented in this chapter employ the formal modeling techniques of modern microeconomic analysis. We omit the mathematical details and state both the models and the conclusions derived from them in a more intuitive form. Readers with training in microeconomics may want to skip the next two sections.
2. A schedule giving price as a function of the number of units sold is called an inverse demand function; the corresponding demand function gives the number of units sold as a function of price.
3. Monopolistic competition refers to competition among sellers of differentiated products (imperfect substitutes) when each of many sellers is so small relative to the market that changes in its price (or other strategic variables) will not be noticed by other firms in the market.

Differentiated products and tastes give rise to downward-sloping demand curves for individual products, as with monopoly, but unrestricted entry in this type of market drives the profits of firms toward zero, as with perfect competition.

4. We ignore here the possibility of perfect price discrimination.
5. The welfare-maximizing price will now be negative rather than zero, because the value of an audience member to an advertiser remains positive even when the value of the program to the marginal audience member falls to zero. In other words, to maximize welfare, it would be necessary to pay some viewers to watch the program who would be unwilling to watch otherwise. The welfare-maximizing viewer payment would equal the value of a viewer to advertisers, so that the disutility of the program to the marginal viewer would just equal the value of the viewer to advertisers. The audience would thereby be expanded to the point at which the sum of viewer and advertiser benefits of the marginal viewer had fallen to zero. Monitoring problems would make such a pricing scheme impossible, however, because viewers could simply leave their sets on to collect their viewing subsidies without actually watching the programs. At best, programs could be made available free.
6. This analysis of pay-supported television has assumed that channel operators are limited to charging a single price to all viewers for access to their programs. If channel operators were able to discriminate among viewers and charge each the full value of viewing a program, the area of viewer benefits under each program's demand curve would be fully reflected in channel revenues, and all programs that contribute to net welfare would be provided. It is possible to reinterpret the demand curves in the Spence-Owen model as representing the aggregate of demands expressed over time through different distribution windows. As noted in Chapter 2, new release windows have opened as new distribution technologies have been developed and have been accepted by viewers. This means that more of the added viewer benefits attributable to individual programs are available to cover program costs. Because of the new technologies, the biases and inefficiencies of a pay-television industry may be decreasing.
7. Under the symmetry assumptions employed below, this ranking is unambiguous.
8. Spence and Owen (1975, 1977) point out that exceptions to this ordering of O, E, and M are possible but not important.
9. A competitive advertiser-supported television industry will always supply enough programs that an additional program would not be able

to generate enough revenue to cover its costs. This places a lower bound on the number of competitive advertiser-supported programs. This number will vary with the demand for advertising.

10. Advertising is becoming more prominent on rented videos. Video cassette distributors now routinely record advertisements for other titles they distribute at the beginnings of cassettes. Occasionally advertisements for other products are included at the beginning or end of the featured program.
11. We are ignoring shopping channels and cable channels such as the Travel Channel that show advertiser-supplied programs promoting the sponsors' products. These services are not numerous, relative to other types of services, and account for only a trivial percentage of the total television audience.
12. This suggests that advertising may be used to support special interest programs with small audiences by forcing viewers to pay a higher "price" for these programs in terms of the amount of program time reserved for advertising.
13. In fact, Wildman and Owen (1985) show that a program will have the same size audience with either advertiser support or viewer payments if (1) the price viewers are willing to pay to eliminate a minute of commercial time is constant and (2) viewers pay only for programs they watch (that is, they do not subscribe to a variety of services and then pick one to watch when the time comes). In this rather special situation, the advertising demand curve is the same as the price demand curve with the vertical axis rescaled by the ratio at which viewers are willing to give up dollars to reduce commercial minutes. Rescaling the vertical axis in this manner has no more effect on the profit-maximizing audience than would a rescaling by measuring price in cents instead of dollars.
14. Even though the Spence-Owen model is a single-period model, demand can be interpreted as generated either by viewers that make exclusive choices among channels or by viewers that have assembled program options prospectively, realizing only one can be viewed at any given time. This also is true for most of the Steiner-type analyses. In this extension of the Spence-Owen model, we assume explicitly that viewers may purchase access to more channels than they can watch at any one time, either because they do not know in advance exactly what programs the channels will be offering or because they know that their preferences concerning content will vary over time. Thus a channel may receive payments from viewers who watch other channels. Because advertisers pay only for the audience that watches

a program, a channel operator has to be concerned with the fraction of time that its subscribers actually are in the audience. The concave (downward) shape of R_a reflects the assumption that the more program time is allocated to commercials, the smaller is the fraction of the channel's subscribers who watch it.

15. According to George L. Schulman, vice president of marketing for the 24-hour movie service Starion, there were 5,000 subscribers to wireless cable services in 1987 and 50,000 to 65,000 in 1988. He anticipated 300,000 subscribers by the end of 1989. See K. R. Clark, "Dishes May Yet Serve TV on Platter," *Chicago Tribune,* December 24, 1989, p. VII-3.
16. Repetition also occurs over much longer periods of time. The broadcast networks repeat episodes from prime-time series during the summer rerun season and syndicated packages of motion pictures usually allow for a film to be shown a number of times over a several-year period.
17. The expected increase in revenue can come from increases in audience and in advertiser payments, increases in charges to viewers or resellers such as cable systems, or both.
18. This is not to deny that weak independent stations may tacitly concede the prime-time period to the stronger network schedules and focus their efforts on the schedules for other dayparts. In aggregate, the resources committed to programming for prime time will still be much larger.
19. Wildman and Lee's analysis is more general than this example. They allow channels' program budgets to reflect the intensity of competition for audience. Expected revenues and program budgets (or costs) fall as the number of competitors increases, but expected revenues fall faster; so channels still switch to schedules of repeated programs.
20. According to Head and Sterling (1990, p. 492), 13 percent of the world's national television broadcasting systems are mixed commercial-public systems, 11 percent are purely commercial, 57 percent are government operated, and the remainder, 19 percent, are systems operated by independent, nonprofit organizations. The number of mixed systems has grown rapidly in recent years as numerous countries have privatized government channels and channels in the hands of other nonprofit programmers.
21. Noam's findings for purely commercial systems largely replicate those of Steiner and Spence-Owen, and his treatment of national television systems that are entirely noncommercial lies outside the scope of this book.
22. In Noam's model, viewers see programs within some range of loca-

tions not too distant from their own location as imperfect, but acceptable, substitutes for each other.

23. See P. Harris, "Congress and Cable Imperil PBS," *Variety*, August 17, 1988, pp. 1, 43, for a more detailed discussion of the plight of public television in the United States near the end of the 1980s.
24. See Litman (1983) and Blumenthal (1988). Litman's regressions show that larger-budget films are more popular, and Zlumenthal shows that theater owners bid more for films with larger production budgets, presumably because they expect them to draw larger audiences.
25. Park (1973, 1975) also allowed program budgets to vary in response to changes in market structure in his studies of the feasibility of a fourth broadcast network. Park was not concerned with program choice, but his model is similar in many respects to that of Wildman and Lee.

5. Network Economics

1. Daily newspapers provide an example of what can happen when economies of scale are more powerful than readers' demand for diversity. There was a time, 50 years or more ago, when most big cities supported a dozen or more daily newspapers. Each newspaper was specialized, not so much by its coverage, but by the political or ethnic orientation of its editorial style. Gradually the demand by advertisers for inexpensive access to newspaper readers, the ability of publishers to serve varied preferences within the pages of a single newspaper, economies of scale in newspaper production, and perhaps a certain homogenization of reader preferences led to consolidation of the daily newspaper industry. Today there are few if any cities large enough to support competing daily newspapers in the long term. (Daily newspaper competition arises from suburban papers, each monopolizing its own sector, rather than from major dailies.) Thus despite the absence of regulatory barriers to entry in the daily newspaper business, few cities have more than one newspaper. Similar economic forces are at work in video broadcasting; if viewer preferences were somewhat different than they apparently are, the dominance of the three broadcast networks might continue despite the reforms and new technologies. But in fact the decline of the broadcast networks is well under way in the early 1990s.
2. "Economic Indicators," *Broadcasting*, June 19, 1989, p. 43.
3. The limitations on commercials during children's programs are 10.5 minutes per hour on weekends and 12.5 minutes on weekdays. One study done prior to the new rules showed that children watched televi-

sion 3.5 hours per day on average, and that 20 percent of their viewing consisted of commercials and other nonprogram material. *New York Times,* April 22, 1991, p. D9.

4. *Gallagher Report,* 36 (23), June 6, 1988, p. 4. Anthony Malara of CBS was quoted by *View Magazine* as estimating 15,000 to 18,000 hours of preemptions for the three networks. "Networkers," *View Magazine,* June 6, 1988, p. 49.
5. These data were graciously supplied by Nielsen Media Research.
6. House of Representatives, Commerce Committee, *Report on Network Broadcasting,* H. R. 1297, 85th Congress, 2d session, 1958 (Barrow Report), p. 207.
7. Affiliates respond to the threat of loss of affiliation to another station in the same market by clearing more network programs (Wildman, 1978).
8. *Time,* October 24, 1983, p. 70; "Turner the Victor in Cable News Battle," *Broadcasting,* October 17, 1983, pp. 27–28.
9. *Time,* October 24, 1983, p. 70.
10. "[T]he most conspicuous reason for their [the networks] current activity in the news arena appears to be the high visibility of Cable News Network and other services as suppliers of national and international news." Josephson, "Webs' Expansion Prodded by Alternative News Services," *Television/Radio Age,* July 26, 1982, p. 33.
11. *Time,* October 24, 1983, p. 70; *Television/Radio Age,* November 26, 1984, pp. 42, 43; Schwartz, "News for Sales," *View,* September 1983, p. 44.
12. "CNN2: Now It's 48 Hours a Day in Cable News Networking," *Broadcasting,* January 4, 1982, pp. 74–75; "Turner the Victor in Cable News Battle," *Broadcasting,* October 17, 1983, p. 28; *Broadcasting,* December 12, 1983, p. 58.
13. Ted Turner identified CNN's head start as a factor contributing to its victory over SNC. His news service prevailed, he explained, because "we started first and we had excellent services." *Broadcasting,* October 17, 1983, p. 27.
14. *Television/Radio Age,* November 26, 1984, pp. 42–43; *View,* September 1983, p. 44.
15. "News Is a Hit on TV's Bottom Line," *New York Times,* September 13, 1989, p. D1.
16. *National Broadcast Co.* v. *U.S.* 319 U.S. 190 (1942). See also Krasnow and Longley (1982).
17. *FCC Rules and Regulations,* Section 73.658, parts (a)–(h).
18. FCC, Office of Network Study (1963, 1965); U.S. House of Representatives (1958).

19. FCC, Network Inquiry Special Staff (1980b), *New Television Networks: Entry, Jurisdiction, Ownership, and Regulation,* vol. 1, pp. 507–518; see also vol. 2, *Background Reports,* pp. 531–539, for a survey of research on the economics of FCC regulation of network program supply.
20. For example, holders of the rights to a series' earnings in off-network syndication will realize a profit only if the series has a successful network run. But the syndication rights are riskier than the network rights because the series that succeed on a network do not always succeed in syndication.
21. On the decline in usage of traditional prime-time entertainment series, see ICF (1983). In this study for the Committee for Prudent Deregulation, ICF found that the number of weekly hours devoted to prime-time series decreased from 60.5 hours on average in the 1965–1970 seasons to 49.2 on average in the decade ending with the 1980–1981 season.
22. ICF's calculation is an understatement, because ICF included miniseries in its count. Part of the decline, of course, is accounted for by the prime-time access rule. See ICF (1983). See also Clift and Greer (1989). It is also likely that the quality or attractiveness of programming will decline, with consequent adverse effects on the public. See Fisher (1985), pp. 290–296.
23. Summer replacements, news (including "60 Minutes"), sports (including "Monday Night Football"), movies (including made-for-television movies), and other programs that were not prime-time entertainment series were excluded. Data were compiled only for series first aired after 1959, and results were calculated only for years beginning with 1963. Before that time the networks themselves and advertisers, rather than Hollywood, supplied a great deal of the programming.
24. A few series from the late 1970s were still on the air in 1990, so their lengths of run were unknown. The longer they last, the greater will be the measured variance of those seasons. In Table 5.11 the series were assumed to run until September 1, 1990. The series still on the air were predominantly prime-time soap operas, such as "Dallas," that do not have significant appeal in domestic syndication and for which the networks were paying virtually all the production costs.
25. Changes in the types and patterns of programs, such as the introduction of "short forms," are probably best viewed as reactions to the costs imposed by the financial interest rule.
26. See ICF (1983). Data in ICF's Exhibit 1-9 at page 1-30 indicate that the number of different producers of prime-time series programs fell

on average after the financial interest and syndication rules went into effect. Similarly, ICF Exhibit 1-10 at page 1-31 shows that concentration among producers of prime-time series increased after the rules were adopted.

27. Sports, news, miniseries, made-for-television movies, and other program forms have replaced much of the prime time once devoted to series.
28. For the studios' position, see Warren-Boulton (1990); for the networks' response, see Crandall (1990e).
29. On ABC's view that the prime-time access rule improved its own competitive position, see Goldenson (1991), pp. 338–339.
30. "Networks Contemplate Cuts in Affiliate Compensation," *Broadcasting,* November 17, 1986, p. 39.
31. "A Surprising Script for NBC," *Business Week,* March 9, 1987, p. 70.
32. *Washington Post,* February 22, 1990, p. E-1; *Broadcasting,* February 26, 1990, p. 27; *Wall Street Journal,* February 22, 1990, p. B-4.
33. *Wall Street Journal,* February 22, 1990, p. B-4.
34. K. Clayton, "Firm Pitching DBS Service Offering PPV," *Electronic Media,* April 23, 1990, p. 4.
35. For a discussion of compression technologies and their possibilities, see "What Cable Could Be in 2000: 500 Channels," *Broadcasting Abroad,* May 1991, pp. 4ff.
36. On the networks' spending on sports broadcast rights, see *Electronic Media,* March 12, 1990, p. 56; *New York Times,* December 3, 1989, section 8, pp. 1, 7. One illustration of the importance, from a business point of view, of the regulatory constraints on efficient network risk bearing can be seen in the recent very large expenditures by the broadcast networks on future sports broadcast rights. The value of these rights has increased for a number of reasons, one of which is that sports broadcast ratings are more predictable than are most entertainment ratings. Consequently the networks place a substantial premium on sports rights with given expected ratings, relative to unproduced entertainment products with equivalent expected ratings. In 1989 CBS paid $1.1 billion for four years of major league baseball and $1 billion for seven years of NCAA basketball tournaments. NBC paid $600 million for four years of NBA games. The three networks together spent $3.6 billion in 1990 for a new NFL contract. The business justification for these expenditures is the relatively low risk of these rights. Early in 1990 the three networks paid nearly $2.8 billion for a NFL rights package.

6. Cable Television

1. For a fascinating parallel, see the description of the British experience with radio broadcasting by wire in Coase (1950), chap. 4.
2. Fortnightly Corp. v. United Artists Television, Inc., 392 U.S. 390 (1968).
3. Midwest Video Corp. v. FCC 406 U.S. 649 (1972).
4. The FCC in 1980 also retained its "syndicated exclusivity rule" that restricted the ability of broadcasters to purchase exclusive local rights to television programs. Like the must carry rule, the syndicated exclusivity rule was abolished nearly a decade later.
5. Technically speaking, the 1984 act eliminated basic rate regulation for all systems subject to competition from at least three broadcast television signals in their service areas. By the late 1980s an overwhelming majority of cable subscribers were served by cable companies satisfying this requirement.
6. FCC, Notice of Inquiry in Mass Media Docket 89-900, released December 29, 1989; Notice of Proposed Rule Making in Mass Media Docket 90–4, released January 22, 1990; Report in Mass Media Docket 89–600, released July 31, 1990 ("Report to Congress").
7. Underscoring this conclusion, federal antitrust agencies declined to take action to stop a 1900 merger between the HA! and the Comedy Channel, and in 1991, despite an intensive investigation, the Federal Trade Commission declined to challenge a proposed merger between the Financial News Network and the Consumer News and Business Channel. These decisions continued a trend established by the Department of Justice in the mid-1980s, when that agency permitted two movie channels to merge. See White (1985).
8. For a discussion of compression technologies and their possibilities, see "What Cable Could Be in 2000: 500 Channels," *Broadcasting Abroad,* May 1991, pp. 4ff.
9. Closely related issues were being litigated by Viacom and Time/Warner in the Southern District of New York in 1990. See complaint filed May 8, 1989, in Viacom International Inc. and Showtime Networks Inc. v. Time Incorporated et al.
10. FCC, Further Notice of Proposed Rule Making, In the Matter of Reexamination of the Effective Competition Standard for the Regulation of Cable Television Basic Service Rates, Mass Media Docket 90-4, released December 31, 1990.
11. The GAO (1989) survey was updated in 1990, as reported in FCC Report in Mass Media Docket 89-600, released July 31, 1990 ("Report

to Congress"), at Appendix F. These data do not alter the conclusions in the text.

12. *Cable TV Programming,* Paul Kagan Associates Inc., November 30, 1988.
13. For example, the average price of a four premium-channel combination decreased 4.9 percent in nominal terms and 10.8 percent in real terms.
14. *The Pay TV Newsletter,* Paul Kagan Associates Inc., July 31, 1989.
15. On the number of subscribers, see *Broadcasting,* March 25, 1985, p. 10; and *Broadcasting,* February 19, 1990, p. 14. On the number of services, see *CableVision,* April 22, 1985, p. 48; *CableVision,* January 15, 1990, p. 74.
16. FCC, Notice of Proposed Rule Making, supra note 6, ¶ 19.
17. FCC, 1985 Staff Study, "Alternative Criteria for Defining Effective Competition: A Statistical Analysis of Small Cable Markets," p. 2.
18. FCC, Notice of Proposed Rule Making, supra note 6, ¶ 21.
19. See "Cable Television Developments," National Cable Television Association, December 1989, p. 5. Viewing shares total more than 100 percent due to multiset usage.
20. Dertouzos and Wildman (1990).
21. Grade B signal contours are themselves a proxy, not an actual measure, of the availability of off-air broadcast signals. Grade B contours are estimates, based on engineering data. Within a Grade B contour, "the quality of picture [is] expected to be satisfactory to the median observer at least 90% of the time for at least 50% of the receiving locations within the contour, in the absence of interfering co-channel and adjacent channel signals." *Television and Cable Factbook,* Stations Volume no. 58, 1990 edition, p. A-13.
22. Within Crandall's data set, it was not possible to distinguish those systems with no over-the-air channels from systems for which this information was missing. Consequently it was not possible to estimate the price difference between systems with no over-the-air channels and systems with one or more over-the-air channels.
23. The finding in Crandall's study that adding additional over-the-air channels past some threshold level does not affect price is similar to the findings by Dertouzos and Wildman. The two studies do differ somewhat, however, in the estimated effect of additional broadcast signals at low levels. In Dertouzos and Wildman's study, the number of Grade B signals did not affect price until five signals were present. A possible explanation for the difference in findings is that the Dertouzos and Wildman sample contained relatively few systems with fewer than five Grade B signals. For instance, only 51 systems in their

sample had fewer than three signals (Dertouzos and Wildman, 1990, table 4, p. 13). As a consequence, any effect of additional signals in this range may have been difficult to detect. Crandall's sample, by contrast, included 257 systems with three or fewer signals.

24. The elasticity of demand is always a negative number. However, we adopt the convention of speaking of the absolute value of elasticity in order to avoid confusion surrounding "increases" in negative numbers.
25. As a general proposition, elasticity of demand may depend on the length of time consumers have to respond to a change in price. Over a long period, a price change may cause a greater quantity change than in a shorter period, because in the long period consumers may have greater flexibility to change their consumption of other related goods. The elasticity of demand for cable is probably not very sensitive to the time period, because changes in consumption of cable and other related goods can be made relatively quickly.
26. For a discussion of the mathematical calculations that are required, and their rationale, see Harris and Simons (1989).
27. This proposition lies behind the Areeda and Turner (1975, p. 716) standard for predatory pricing used in antitrust analysis.
28. These results were reported in *Channels,* October 22, 1990, p. 72.
29. Wildman and Owen (1985) show that there are circumstances in which single-channel competitors may be more effective than a second multichannel competitor in forcing down the price of an incumbent multichannel service.
30. It must be admitted, however, that it is unclear which antitrust enforcement policy results in the lower subscriber prices being charged by incumbents. If entrants know that they will be unable to merge with incumbents, is entry more or less likely? If it is less likely, incumbents need not rely so heavily on a low, entry-deterring subscription price.
31. *CableVision,* January 15, 1990, p. 74. Crandall assumed total national subscribers to be approximately 52.3 million. *Broadcasting,* February 19, 1990, p. 14.
32. However, although multichannel networks (such as HBO/Cinemax or Showtime/The Movie Channel/MTV) do not appear to offer overwhelming production cost advantages over single-channel networks, there may well be significant economies in marketing bundles of channels, as discussed earlier in this chapter.
33. Of course, there are also local markets for certain programs such as local news and local sports.

34. For a general discussion of monopsony power, see Carlton and Perloff (1990), pp. 114–117.
35. For a discussion of multichannel video competition, see Chapter 4.
36. One can also think about this proposition from the more usual perspective of seller market power. Cable systems can be regarded as selling services to national video program providers. (These services include assembling a popular bundle of video offerings, local promotion, distribution to subscribers, and billing.) Services provided by "monopoly" cable system A would be a complement, not a substitute, for services provided by "monopoly" cable system B. The video provider would buy both types of service together to produce delivered programs. Furthermore, if the price of service from A goes up, then the provider will wish to produce less product and so will buy less service from B. Although two merging firms that each monopolize products that are (imperfect) substitutes will raise prices, merging firms that produce complementary products will lower prices. This happens because raising the price of a product lowers the demand for its complements, and the merger will internalize and adjust for this externality. Thus, unlike the case in which competitors merge and raise prices, injuring consumers, a merger between producers of complementary products may (and frequently will) lower prices and enhance consumer welfare. For a general discussion of this phenomenon, see Bulow, Geanakopolos, and Klemperer (1985).
37. Ordover, Sykes, and Willig (1985). In the case of pay networks, however, some researchers have found evidence that HBO's competitors may face some problems on ATC systems. ATC and HBO are owned by Time/Warner; ATC is the second largest cable MSO.
38. Current national subscribers are reported in *CableVision,* January 15, 1990. Nick at Nite, FNN Sports, and C-Span II were excluded because of insufficient data.
39. For example, would constitutional guarantees be able to withstand the public perception that cable operators had substantial economic power? Even less compelling rationales for content regulation, such as the "scarcity" argument that the Supreme Court found sufficient to uphold the FCC's so-called fairness doctrine, have been used to justify restrictions on the First Amendment rights of broadcasters. Red Lion Broadcasting Co. v. FCC, 395 U.S. 367 (1969).
40. See, for example, Morrison and Winston (1986); Winston et al. (1990). For a survey of the effects of economic regulation, see Joskow and Rose (1989).
41. See the materials in FCC Common Carrier Docket 87-313.
42. California, for example, has adopted limited price caps for local

exchange carriers. Decision 89-10-031, Public Utilities Commission of the State of California, October 12, 1989.

43. See Joskow and Rose (1989) for a concise summary of some of this literature.
44. The FCC's lengthy efforts to disentangle AT&T's service costs during the 1962–1981 period stand as testimony to this problem.
45. An excellent summary of these effects may be found in the FCC's Further Notice of Proposed Rule Making, Docket 87-313, released May 12, 1988.
46. We assume here that local stations would acquire the rights to license their programs to local cable systems. It is perfectly plausible, however, that broadcast networks would acquire these rights and license cable systems directly for the network portion of local station broadcasts.

7. Advanced Television

1. For details, see "Current Standards" and "Proposed Standards," below.
2. See Besen and Johnson's (1986) case histories of standards selection in various broadcast industries.
3. FCC, Tentative Decision and Further Notice of Inquiry, In the Matter of Advanced Television Systems and Their Impact on the Existing Television Broadcast Service, Mass Media Docket 87-286, released September 1, 1988.
4. In the discussion that follows, we assume that a technology is defined by production standards. Thus in some contexts "standards" and "technology" may be used interchangeably.
5. For reveiws of this literature, see Besen and Johnson (1986) and David (1986). For a discussion of standard setting in television, see the 1988 Report of Working Party 5 to the FCC Advisory Committee on Advanced Television Service, section 8. Appendix 3 of this report is a bibliography of the standards literature.
6. Many readers will be more familiar with analyses of negative externalities, such as pollution and freeway congestion. The typical problem with negative externalities is that in maximizing their personal welfare, individual decisionmakers consume too much of products and services characterized by negative externalities because they ignore the harmful effects of their actions on others. The efficiency problem with positive externalities is more likely to be underconsumption. Individuals consume too little of those products and services for which there are positive externalities because some of the benefits of their actions

are realized by others. Individual pursuit of gain leads to efficiency through the operation of the invisible hand only if all the consequences of interactions among individuals are internalized in market transactions.

7. For one of the earliest and most lucid presentations of the economics of standard setting, see Arthur (1983).
8. The extent to which lower costs are passed on to consumers through price cuts will depend in part on the vigor of competition in the industry. In a competitive industry, all cost savings will eventually be passed on. Even a monopolist will usually be compelled by self-interest to lower prices in response to a cost reduction.
9. Some users adopt new technologies before others. Initial users may anticipate the greatest benefits from a new technology (because it is particularly suited to their operations or because they are close to the point at which they would normally replace equipment based on an older technology). Or, because of professional circumstances or personal interests, early adopters may be those who are the most informed about technological developments. They also may be natural "plungers" who are inclined to take risks rather than pursue a "wait and see" strategy with respect to new technologies.
10. Each entry in Table 7.1 gives the value to A of being able to use a particular technology depending on the technology used by B when the option is using neither of the two technologies.
11. Arthur (1983) was the first to represent diagrammatically the effects of the size of the user base on the benefits of a technology to individual users in this manner.
12. Farrell and Saloner (1985) have shown that if each user believes the new technology is superior to the old technology, and is fully confident that others will switch if he or she does, then the switch will be made.
13. Diagrammatically, N'' is determined by finding the unique horizontal line for which the intersection with the two technology benefits schedules has the number of users of the new technology plus the number of users of the old technology equal to N^*.
14. Farrell and Saloner (1985) prove, through repeated application of this logic to all N^* users, that a single user may induce all to switch, even when some would prefer that all stick with the old technology. This would be the case if a user preferred that all switch and knew that by switching it would place a second user in the position of both preferring that all users switch and being able to place a third user in exactly the same position with respect to a fourth user, and so on. The first user would then be able to initiate a chain reaction of switching to the new technology that eventually would induce all to switch. Note,

however, that this proof relies on the unrealistic assumption that each user has perfect information about the preferences of all users that have not switched.

15. Unlike the simple industries portrayed in this discussion, most industries have multiple stages (for example, design, manufacturing, distribution, retailing, production, transmission, broadcasting). It is the largest firms at the most concentrated stage of production that have the most influence on the selection of standards (Besen and Johnson, 1986).
16. For example, patented technologies would have an advantage relative to technologies that are in the public domain, because their sponsors could capture more of the benefits of their adoption.
17. Shared expectations of computer users and software writers that the IBM PC would dominate the personal computer market is sometimes offered as an explanation for the rapid emergence of the IBM PC operating system (MS-DOS) as a de facto standard for personal computers. The IBM PC was a rather late entrant to the personal computer market, and it offered few if any technological advances over personal computers already on the market. Yet within a few years of its introduction, most personal computers sold were IBM PCs or PC clones, and most of the surviving manufacturers of personal computers with other operating systems had either switched to the IBM operating system or had adapted them so they could run its software. Prior to the IBM PC, there existed a large number of personal computers utilizing a variety of operating systems of varying degrees of compatibility, none of which had established a position of dominance. The shift by users to the IBM PC may have been motivated by a desire to avoid the risks of purchasing other personal computers, the long-term success of which seemed far less certain.
18. Because each user selects a standard on the basis of the personal benefits involved and ignores benefits that might be conferred on users of other standards not adopted, too many standards may survive for economic efficiency. This would be the case if there were many standards, each with a small number of users, when all would be better off with a single standard. The inability of new standards to establish themselves when older standards are locked in may, of course, prevent the emergence of technologies that might serve the interests of smaller, specialized groups of users. Thus equilibria with too many and too few standards are both possible.
19. Many examples of engineered compatibility exist. Various peripheral devices make it possible to view programs transmitted by cable, satellite, and recorded cassettes and disks on a standard NTSC television

receiver. Manufacturers of computer printers design them so they can be used with different types of computers. Gateway devices are designated to connect private local area communication networks (LANs) to the public telephone network, even though different communication technologies generally are involved.

20. This function is also served by the many voluntary industry standards associations. For an interesting study of the activities of private, voluntary standards organizations, see Weiss and Sirbu (1988).
21. See the comments of William F. Schreiber (November 30, 1988) before the Federal Communications Commission, In the Matter of Advanced Television Systems and Their Impact on the Existing Television Service, Mass Media Docket 87-286.
22. See 2 F.C.C. 2d 17 (1987) (No. 87-268) for a discussion of the many aspects of television transmission governed by the NTSC standards.
23. Other improvements will be generated by engineering changes that will reduce or eliminate ghosts and by techniques for improving luminance and chrominance resolution. Many of these improvements can be captured with IDTV receivers, and others can be generated with slight modification of the NTSC transmission standards or by preprocessing production signals prior to transmission.
24. This definition was used by the FCC in tentative decision FCC release 88-288, 37462 (1988), Docket 87-268.
25. Flexible receivers, which may or may not be of the OAR variety, that can process different kinds of signals may be required to allow each medium to employ standards best suited to its particular technology.
26. "NBC Unveils New HDTV Standard," *Broadcasting,* October 17, 1989, p. 31.
27. 2 F.C.C. 2d 17 (1987) (No. 87-268).
28. Of course, ATV productions could not be viewed at ATV quality with NTSC receivers, and NTSC productions would not provide ATV quality if viewed with ATV receivers.
29. There are alternative techniques for resolving the aspect ratio problem. The side portions of ATV pictures could be compressed for use with NTSC, which would result in some picture distortion. Or programs could be produced with the NTSC aspect ratios and then the tops or bottoms of the pictures could be cropped for ATV broadcasts. Both of these solutions would still encounter difficulties similar to those discussed in the text.
30. By equating program supply with program production, we are assuming that a firm that takes responsibility for supplying programs to distributors also bears the financial responsibility for ensuring that programs are produced according to the appropriate standards.

31. CBS did sponsor an earlier color television standard that failed to achieve FCC approval.
32. With current receiver technology, a large-screen ATV receiver will occupy as much space as a standard household refrigerator.
33. For reports of other ATV adoption forecasts, see Nathan Associates (1989). None of the major forecasts has explicitly considered the relationship between NTSC compatibility and the rate of ATV adoption. However, they all are extrapolations from adoption histories of consumer electronics goods, such as the VCR, the compact disc player, and color television, that were compatible with existing technologies when they were introduced.
34. *Broadcasting,* April 9, 1990, pp. 40–41; April 16, 1990, p. 43; April 30, 1990, p. 35.
35. Robert D. Reischauer, the director of the Congressional Budget Office, made this point before the Senate Committee on Governmental Affairs on August 1, 1989.
36. As we have drawn these schedules, the maximum benefits attainable with the compatible systems are somewhat less than the benefits promised by their incompatible counterparts. This reflects the possibility that compatibility requirements may limit the quality achievable with the compatible systems to levels below what would be possible if compatibility were not a concern. However, compromises on system quality may increase the likelihood of commercial success if they increase compatibility with a previously established system (see Figure 7.4).
37. Foreign ownership of U.S. broadcast stations is restricted by law, and Democratic and Republican administrations alike have tried to influence broadcast coverage of foreign affairs. Nevertheless, viewed against the backdrop of the ongoing debate concerning the domestic television industry, issues related to foreign policy must be considered peripheral at best.
38. See C. Kippo, "In Search for Visual Perfection, Budget Referees HD vs. Film Fight," *Variety,* October 5, 1988, p. 238; "Television: The Challenge of the Future," address by Joseph A. Flaherty to SMPTE Winter Conference, February 7, 1987; R. Stow, "The Economics of High Definition Television Production," unpublished paper, March 9, 1987; R. Stow, "HDTV—Making It Happen," paper delivered at Probe Research Inc. HDTV Symposium, November 16, 1988, p. 7.
39. Conversion of ATV programs produced under an 1125/60 production standard for broadcast on conventional (NTSC) American television results in a higher-quality product than if the program had been produced by 35-millimeter film and converted to NTSC. See Stow,

"HDTV," p. 4. In addition, the cost of this conversion is low. The equipment needed to convert 1125/60 tape to 525/59.94 costs about $120,000. See also P. Robert, M. Lamnabhi, and J. J. Lhnillier, "Advanced High Definition 50 to 60 Hz Standards Conversion," *SMPTE Journal* 98 (June 1989): 420–424.

40. "Bottom-Line Impact of HDTV," *Broadcasting,* September 19, 1988, pp. 74, 76.
41. This is apparently the view of Thomson Consumer Electronics, Inc., the French-owned supplier of RCA television sets, which is actively involved with NBC in the development of an American standard. See D. Joseph Donahue, Vice President of Thomson Consumer Electronics, Inc., "HDTV: The Real Issue Is American Technology and Jobs," unpublished paper, p. 6.
42. For example, the one remaining U.S.-owned producer of television sets, Zenith, has assembly facilities in Mexico. See National Telecommunications and Information Administration, 1988, p. 622.
43. Donahue, "HDTV," p. 3.
44. National Telecommunications and Information Administration, 1988, pp. 622, 623.
45. Nathan Associates, 1989, pp. 3, 19. In the Nathan study, compatibility means that whatever ATV technology is adopted, NTSC television sets will not be made obsolete. This would require that the ATV distribution standard be compatible with the NTSC distribution standard—the same requirement that the FCC has decided to impose. See Tentative Decision and Further Notice of Inquiry, Mass Media Docket 87-268 (September 1, 1988), ¶¶ 4, 125–126.
46. Cellular radio technology and standards were developed in the United States, but imports still account for a large proportion of cellular equipment used in the United States and would account for more were it not for trade barriers. See Besen and Johnson (1986), p. 134.
47. "Sony Corp. Would Bite the Bullet If European HD System Adopted," *Variety,* October 5, 1988, p. 92.
48. With rare exceptions, television markets receive considerably less than the full complement of six VHF stations because available channels are divided among neighboring urban areas.

Index